Roar of the Ages:

The Life Story of Drakon, the Mighty Tyrannosaurus

By

Yoosuf Bhutta

Roar of the Ages: The Life Story of Drakon, the Mighty Tyrannosaurus

Yoosuf Bhutta

Published by Yoosuf Bhutta, 2025.

ROAR OF THE AGES: THE LIFE STORY OF DRAKON, THE MIGHTY TYRANNOSAURUS

First edition. January 15, 2025.

ISBN: 979-8230386568

Written by Yoosuf Bhutta.

Table of Contents

"In the shadow of ancient giants, we uncover the echoes of a world long gone, where power and survival were etched into the bones of time."— Unknown

Introduction

The World of Dinosaurs

Imagine a time when Earth's continents were still forming, and a warm, humid climate fostered life on an epic scale. This was the Mesozoic Era, a period stretching over 180 million years, divided into the Triassic, Jurassic, and Cretaceous periods. It was a world filled with giant ferns, coniferous trees, and vast, unspoiled landscapes, where creatures unlike anything we see today thrived. From swift raptors that hunted in packs to towering sauropods that grazed on treetops, the planet was ruled by an array of reptiles, each adapted to its environment in extraordinary ways.

At the pinnacle of this world stood the dinosaurs, the undisputed rulers of land, air, and sea. These creatures weren't just larger-than-life; they were marvels of evolution, showcasing nature's ability to adapt, survive, and thrive. And towering above all, both in stature and in reputation, was Drakon, the mighty Tyrannosaurus rex. His name alone evoked fear and respect across the ancient landscape, and his reign as the apex predator was a testament to his immense strength, cunning, and sheer survival instincts. His roar could silence the thick, primeval forests, reverberating across the plains as a warning to all.

This tale is not merely about Drakon's life but the dynamic world he inhabited—a world teeming with volcanic activity, shifting tectonic plates, and evolving ecosystems. The Earth during this period was far more volatile and untamed than today's world, and this instability shaped the lives of the creatures that roamed its surface.

The Mesozoic Era is often called the "Age of Reptiles," but it was also an age of dramatic environmental shifts and intense competition for survival. From the scorching deserts to lush swamps, dinosaurs adapted to every conceivable environment, their evolution constantly pushing the boundaries of what life could achieve. It was in this brutal, beautiful world that Drakon, like his fellow predators, learned to survive and dominate.

Through this book, you will embark on a thrilling journey through the life of Drakon—from his early days as a vulnerable hatchling to his rise as the most feared creature of his time. But you'll also witness the incredible variety of life around him, the ever-changing landscapes, and the intricate web of predator and prey. This is not just a story of survival, but a story of the Earth itself, at a time when nature's forces were at their most magnificent and unforgiving.

So, as we begin this journey, take a step back in time—over 65 million years ago—and prepare to witness the majesty of the dinosaurs, a world dominated by giants, and the legendary life of Drakon, the mighty Tyrannosaurus rex.

Quick Facts:

❖ Fossilization occurs when an animal is buried quickly, and minerals replace its bones over millions of years.

❖ Paleontologists use tools like brushes, chisels, and even dental picks to excavate delicate fossils.

❖ The term "dinosaur" was coined in 1842 by Sir Richard Owen, meaning "terrible lizard."

❖ CT scans and 3D printing allow scientists to study fossils without damaging them, revealing new insights about dinosaur anatomy.

Did You Know?
Tyrannosaurus rex had one of the most powerful bites in history. T. rex could exert a bite force of over 12,000 pounds per square inch—enough to crush the bones of its prey with ease!

Dedication

To the wonders of the natural world, both ancient and modern.

To every curious mind that seeks to understand the marvels of Earth's past.

And to the tireless explorers, scientists, and educators who continue to reveal the secrets of life long gone.

This story is for the dreamers who look back to a time of giants and for the stewards of our planet today, who recognize the importance of preserving what we have for future generations.

PART-One

Chapter 1: Overview of the World

1.1 Overview of the Mesozoic Era

1.2 The Climate and Geography of Earth

1.3 The Age of Reptiles

1.4 Tyrannosaurus rex as the apex predator

Chapter 1:

1.1 Overview of the Mesozoic Era

The Mesozoic Era spanned an incredible 180 million years, a vast stretch of time divided into three periods: the Triassic, Jurassic, and Cretaceous. To give you a sense of just how long ago this was, humans have only been around for a tiny fraction of that time—about 300,000 years. During the Mesozoic, dinosaurs ruled the Earth, but it wasn't always so.

The Triassic Period (around 252 to 201 million years ago) was when dinosaurs first appeared, small and unassuming. At that time, the Earth was dominated by reptiles, and the landmasses were clumped together in one giant supercontinent called Pangaea.

- Then came the Jurassic Period (201 to 145 million years ago), where dinosaurs flourished. It was during this time that the world began to split into the continents we recognize today. Some of the most iconic dinosaurs, like the Brachiosaurus and Stegosaurus, roamed the lush landscapes.

Finally, the ***Cretaceous Period*** (145 to 66 million years ago)—this was the age of Drakon, the apex predator of his time. By now, dinosaurs had become the undisputed rulers of the land, evolving into an array of remarkable forms, from towering herbivores to swift, deadly carnivores like the Tyrannosaurus rex.

It was a time when life exploded in diversity, and dinosaurs became the masters of survival. But it wasn't just their size or strength that made them dominant—it was their ability to adapt to the Earth's ever-changing conditions.

1.2 The Climate and Geography of Earth

The Earth of Drakon's time was a planet of extremes. It was a place of vast forests, inland seas, and towering mountains. The climate was significantly warmer than it is today, and there were no polar ice caps. Imagine a world where lush green forests stretched for miles, broken only by the massive rivers that snaked their way through the landscape. The atmosphere was thick with humidity, creating a tropical paradise where plants grew tall and dense.

Drakon's world was vibrant and teeming with life. Conifer trees, ancient ancestors of today's pines, stretched towards the sky. Ferns blanketed the forest floor, providing a rich underbrush where smaller animals could hide. The air smelled of damp earth, the result of frequent rains and the ever-present moisture in the air. It was a landscape shaped by volcanic activity and continental drift, as the Earth's plates shifted beneath the surface, slowly but constantly transforming the world above.

During the Cretaceous Period, the Earth was in the process of breaking up from the supercontinent Pangaea. The land that Drakon roamed was part of what is now North America, but at the time, large portions of the continent were submerged under shallow inland seas. These seas were teeming with life, from giant marine reptiles to fish and other sea creatures that were unlike anything we see today.

The warm, humid climate and the abundance of plant life created the perfect environment for herbivorous dinosaurs to thrive, which in turn, provided plenty of prey for Drakon. It was a world where survival depended on size, strength, and adaptability, and Drakon had all three in spades.

1.3 The Age of Reptiles

The Mesozoic Era was not just the Age of Dinosaurs—it was the Age of Reptiles. While dinosaurs ruled the land, the seas were dominated by massive marine reptiles like Mosasaurus and Plesiosaurs, and the skies were filled with flying reptiles like Pteranodon. But it was the dinosaurs, the mighty land-dwellers, that truly defined this age.

Among all the dinosaurs that lived, few have captured the imagination of people as much as the Tyrannosaurus rex. Standing over 12 feet tall at the hips and measuring up to 40 feet in length, the T. rex was a marvel of evolutionary engineering. With teeth that could grow up to 8 inches long, it had the strongest bite force of any land animal that ever lived. Drakon, the Mighty T. rex, was not just a predator—he was an apex predator, sitting at the very top of the food chain.

Drakon's senses were finely tuned for hunting. His keen eyesight allowed him to spot prey from great distances, and his powerful legs gave him bursts of speed that made him deadly in close combat. But it was his sense of smell that truly made him the terror of his domain. With a nasal cavity larger than that of any other dinosaur, Drakon could smell his prey from miles away, tracking them through dense forests and across plains.

His environment was filled with other dinosaurs, like the three-horned Triceratops, the massive Alamosaurus, and the agile Ornithomimids. These creatures lived in constant fear of Drakon, who hunted alone, relying on his sheer strength, intelligence, and fearsome reputation.

The Age of Reptiles was a time of incredible variety and grandeur. From the towering Sauropods that grazed peacefully to the deadly Velociraptors that hunted in packs, the land was alive with creatures that had adapted to every possible niche. But in Drakon's world, being the largest and strongest wasn't always enough—survival required cunning, speed, and the ability to adapt to an ever-changing environment.

1.4 Tyrannosaurus Rex as the Apex Predator

The "*Tyrannosaurus rex*", often dubbed the "king of the dinosaurs," was the apex predator of its time, standing at the pinnacle of the food chain during the late Cretaceous period, around 68 to 66 million years ago. Measuring up to 40 feet (12 meters) in length and weighing an estimated 9 tons, ***T. rex*** was an awe-inspiring beast whose dominance was unrivalled in the ecosystems it inhabited.

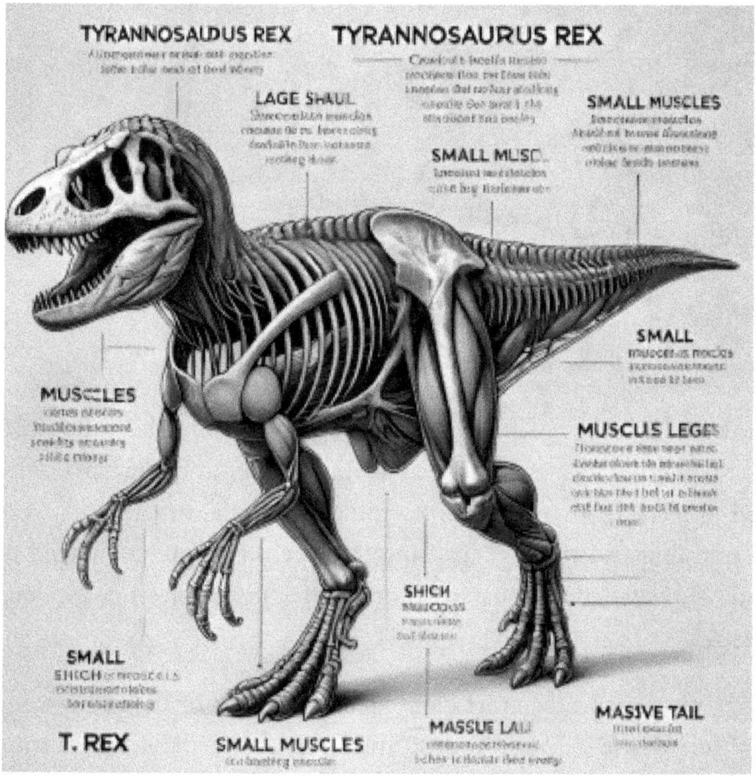

Anatomy Built for Power and Efficiency

T. rex was equipped with features that made it an ideal predator. Its massive skull, which could reach up to 5 feet in length, housed powerful jaw muscles capable of delivering a bite force of nearly 12,800 pounds—the strongest bite of any terrestrial animal ever recorded. This incredible bite force allowed ***T. rex*** to crush bone, an ability that gave it access to the nutrient-rich marrow inside its prey's skeletons.

Its teeth, some of which were 12 inches long, were not only sharp but also serrated like steak knives, designed to tear through flesh and shatter bones. Unlike the slicing teeth of many predators today, ***T. rex*** had teeth that were built for both gripping and tearing, ensuring that even if its prey managed to escape, it would likely succumb to massive blood loss or infection from the deep wounds inflicted.

Though its arms were comically short compared to the rest of its body—only about 3 feet in length—they were robust and muscular, capable of exerting significant force. While these arms were not used for grappling prey, their specific function remains a subject of debate among paleontologists, with some theories suggesting they were used for mating or helping the animal rise from a prone position.

A Predator with Superior Senses

The sensory capabilities of ***T. rex*** were also finely tuned for hunting. Recent studies suggest that its eyesight was far superior to that of most modern-day predators, with binocular vision allowing it to judge distances with high accuracy. This would have been a crucial advantage when chasing down prey or stalking it through the Cretaceous forests.

Additionally, ***T. rex*** had an extraordinary sense of smell, possibly one of the best among dinosaurs. This acute olfactory sense helped it detect prey over long distances, making it an excellent scavenger as well as an active hunter. Whether hunting or waiting for an animal to die of natural causes, its keen nose would lead it to fresh kills.

Its hearing was equally impressive, particularly in detecting low-frequency sounds. This would have allowed it to hear the footsteps or vocalizations of other large animals over considerable distances. Combined with its sense of smell, ***T. rex*** was an efficient, opportunistic predator.

Hunting Techniques and Predatory Behavior

As a highly intelligent predator, ***T. rex*** likely employed strategic hunting methods. Some paleontologists suggest that it may have been a pursuit predator, using its muscular legs to run at speeds of up to 20 miles per hour in short bursts. Its large, muscular legs were ideal for covering ground quickly, helping it close in on prey like hadrosaurs, ceratopsians, and other herbivorous dinosaurs of the time.

However, ***T. rex*** may not have relied solely on speed. Its formidable size, combined with the element of surprise, would have made it a lethal ambush predator. Hiding in dense foliage or stalking through the cover of forests, ***T. rex*** could strike when its prey was least expecting it. Once engaged, the sheer force of its jaws, along with the ability to deliver fatal bites, meant that most animals had little chance of survival.

There is also evidence to suggest that ***T. rex*** was not above scavenging. Its exceptional sense of smell could lead it to carcasses from miles away, where it would then use its powerful jaws to crack open bones and extract every possible nutrient from the remains. This opportunistic behavior, blending active hunting with scavenging, allowed *T. rex* to maintain its apex status with minimal energy expenditure when necessary.

Did You Know?

Some dinosaurs were as fast as cars!

The speedy dinosaur ***Ornithomimus*** could run at speeds of up to 40 miles per hour, making it one of the fastest dinosaurs.

Apex Predator in a Competitive World

Despite its dominance, ***T. rex*** lived in a world filled with other formidable creatures. While adult ***T. rex*** had few natural predators, younger individuals may have been vulnerable to attacks by other carnivorous dinosaurs, like "Dromaeosaurs" or smaller theropods. To survive the competitive environment, ***T. rex*** juveniles likely honed their skills from a young age, gradually ascending to their role as apex predators as they grew larger and stronger.

In this dynamic, ever-shifting environment, **T. rex** adapted remarkably well to its ecological niche. It thrived as a solitary hunter, relying on its extraordinary physical and sensory capabilities to dominate its territory. Fossil evidence of healed bite wounds and broken bones suggests that even confrontations between adult ***T. rex*** individuals occurred, likely in battles over territory, food, or mating rights.

The End of Tyrannosaurus Rex

Though ***T. rex*** was at the peak of its evolutionary prowess, its reign was abruptly ended by the mass extinction event 66 million years ago, likely caused by a combination of an asteroid impact and volcanic activity. This cataclysm brought about drastic changes to the Earth's climate and environment, wiping out nearly all large dinosaurs, including ***T. rex***.

Nevertheless, ***Tyrannosaurus rex*** remains one of the most iconic and well-known dinosaurs, embodying the might and terror of prehistoric predators. Its legacy continues to captivate the human imagination, reminding us of a time when nature's balance was dictated by the rule of the apex predator, ***T. rex***.

In this thrilling age of giants, Drakon stood as the king of the dinosaurs, feared and respected by all. The Mesozoic Era was a time of great transformation, with creatures that pushed the limits of what life on Earth could be. It was an era where nature's creativity was at its peak, giving birth to species that would inspire awe and wonder for millions of years to come.

This is the world in which Drakon lived—a world where survival was a daily battle, and where one roar could silence an entire forest. As we journey through his life, from his hatching to his final days, we will see not only the incredible power of dinosaurs but also the fragility of life on Earth, as forces beyond their control began to shape the end of their reign.

The stage is set. The world is ready. Now, let's begin the life story of Drakon, the mighty Tyrannosaurus rex.

Quick Facts:

- ❖ Carnivores like Tyrannosaurus rex were apex predators, while herbivores like Triceratops had built-in defenses such as horns.
- ❖ Velociraptors and other small theropods likely had feathers, showing a link between dinosaurs and birds.
- ❖ Marine reptiles like Mosasaurus and flying reptiles like Pteranodon ruled the oceans and skies but were not technically dinosaurs.

Did You Know?

The longest dinosaur was over 100 feet long!
The ***Argentinosaurus***, a giant sauropod, could grow to over 100 feet long—about the length of a tennis court!

Chapter 2: The Birth of Drakon

Chapter 2: The Birth of Drakon

2.1 Hatching in a World of Giants: *Drakon's early life, from egg to hatchling*

In a vast, ancient forest filled with the sounds of rustling leaves, chirping insects, and the occasional distant roar of a predator, a new life was about to begin. Beneath the thick canopy of trees, in a hidden nest of twigs, leaves, and mud, a cluster of large leathery eggs lay nestled in the soft earth. One of these eggs, about the size of a soccer ball, began to shake gently. Cracks appeared on its surface, and from within, a small, clawed foot pushed through the shell.

This was the moment that Drakon, the mighty Tyrannosaurus rex, entered the world. But unlike the towering, fearsome predator he would one day become, Drakon was now a vulnerable hatchling, surrounded by the shadows of giants.

In Drakon's world, being born was just the first challenge. The forest he emerged into was filled with danger at every turn. Even the mightiest creatures began life as fragile infants, and survival depended on speed, stealth, and a bit of luck. Drakon's nest, located near the edge of a dense forest in what is now modern-day North America, was carefully chosen by his mother—safely hidden from the wandering eyes of predators like Dromaeosaurs or even other Tyrannosaurs, who might view a clutch of eggs as a convenient meal.

For Drakon, hatching was a struggle, but it was only the beginning. His tiny, sharp claws and egg tooth (a small, temporary tooth used to break out of the egg) helped him push through the leathery surface. With one final effort, he emerged, blinking in the dappled sunlight of the ancient world, barely the size of a chicken, yet destined to be one of the largest and most feared predators to ever roam the Earth.

2.2 Appearance and Early Life: *Describes Drakon's initial size, colors, and behavior*

As a newborn, Drakon was hardly the imposing creature history would later remember. Instead, he was a small, scaly hatchling, his body covered in a soft, downy layer of proto-feathers that many young dinosaurs had, likely for warmth. His skin was a pale brownish-green, an ideal camouflage that allowed him to blend into the forest floor as he darted between patches of undergrowth and fern beds. Nature had equipped him with a color scheme that made him almost invisible against the forest's earthy tones, a necessary adaptation for his survival.

His head was disproportionately large compared to his body, with tiny, almost comical arms that would eventually grow into the iconic two-fingered limbs of the Tyrannosaurus rex. Though his teeth were small and not yet capable of the devastating bite of an adult, they were sharp enough to tear into small prey. His early diet consisted of insects, lizards, and even small mammals that scurried along the forest floor, unaware that they were soon to be the first meals of a predator that would one day become an apex hunter.

In his first few weeks, Drakon was entirely dependent on the protection of his mother. She was never far, her towering form casting a long shadow over the nest. Her senses were keen, her maternal instincts sharp. Any predator that ventured too close would be met with a terrifying roar and the deadly force of her jaws. In these early days, Drakon was safe in the presence of his mother, but as with all Tyrannosaurs, he would soon have to fend for himself.

Drakon's siblings, hatched alongside him, also scrambled around the nest, competing for the same food and space. Though they were born together, their lives would soon diverge as they ventured into the dangerous world of the Cretaceous on their own. Few would survive, and fewer still would grow to be as powerful as Drakon. The odds were against him, but nature had gifted him with the instincts and tools he needed to thrive in this untamed land.

2.3 Early Challenges: *Details survival struggles as a young dinosaur*

The first few months of Drakon's life were filled with challenges that tested his resilience. The Cretaceous period was not an easy time for young dinosaurs, especially for a species like the Tyrannosaurus rex. Although Drakon's mother fiercely protected her nest, there were countless dangers lurking in the shadows, each waiting for the moment to strike when her back was turned.

One of Drakon's greatest threats came from Dromaeosaurs, smaller but highly intelligent predators that hunted in packs. These creatures, known to some as "raptors," were quick, agile, and relentless. While they posed no threat to an adult Tyrannosaurus, they were dangerous adversaries for a hatchling like Drakon. On more than one occasion, his mother had to chase away these predators, her roars echoing through the forest, sending the smaller carnivores scattering into the trees.

Then there were the natural challenges. The Cretaceous climate, though warm and hospitable, could be unpredictable. Sudden storms would sweep through the forests, turning the ground into treacherous mud pits and causing flash floods that could wash away everything in their path. Drakon learned early on to seek higher ground, following his mother's lead to avoid the danger posed by rising waters.

Drakon also had to contend with his own kind. Young Tyrannosaurs were not immune to sibling rivalry, and as he and his siblings grew, their competition for food became fierce. In the animal kingdom, survival was often a brutal contest, and Drakon quickly learned that he would have to fight for his share of food. His siblings were not his friends—they were rivals. If he wanted to survive, he had to be faster, stronger, and more cunning than them.

Despite these challenges, Drakon's instincts were sharp, and his body grew stronger with each passing day. He began to venture further from the nest, exploring the dense foliage and learning the lay of the land. Every step was a lesson in survival, every encounter a test of his wits and reflexes. His mother still watched from afar, but Drakon knew that one day soon, he would have to leave her protection and face the dangers of the world on his own.

The early challenges of life had shaped Drakon into a fierce survivor. Each day brought new obstacles, but with them came new opportunities to learn, grow, and adapt. Soon, the little hatchling would leave behind his vulnerable past and step into the next phase of his life, where he would begin his ascent to the top of the food chain.

Drakon's journey was just beginning, but the trials of his youth had prepared him well for what lay ahead. As he grew, his world expanded, and with it, the challenges he would face would become greater. But so too would his strength, his speed, and his hunger for the hunt.

Drakon was no longer just another hatchling in a nest. He was destined to become a king.

Quick Facts:

❖ Birds are the direct descendants of theropod dinosaurs, like Archaeopteryx, the "first bird."

❖ Feathers evolved in dinosaurs before they were used for flight and likely served for insulation or display.

❖ Modern reptiles like crocodiles and turtles share evolutionary traits with their ancient dinosaur relatives.

Did You Know?

Velociraptors were covered in feathers!
Unlike the scaly predators we often see in movies, Velociraptors had feathers and were more like large, predatory birds.

Chapter 3: Growing Up: Drakon's Juvenile Years

3.1 Diet and Feeding

3.2 The Ecosystem and Other Dinosaurs

Chapter 3: Growing Up: Drakon's Juvenile Years

3.1 Diet and Feeding: *How Drakon's diet changed as he grew*

As Drakon grew from a vulnerable hatchling into a more robust juvenile, his world—and his appetite—expanded dramatically. No longer content with the small insects and lizards that had sustained him in his early days, Drakon's diet evolved in step with his increasing size and strength. With each passing day, his jaws became more powerful, his teeth sharper, and his hunger more voracious.

In the first months of his juvenile life, Drakon began to explore beyond the protective shadow of his mother. His newfound independence meant he was now responsible for his own meals, and the forest provided a bountiful but challenging hunting ground. With his keen eyesight and sharp sense of smell, Drakon stalked through the dense undergrowth, hunting for larger prey than he'd ever attempted before.

During this phase of his life, Drakon feasted primarily on small to medium-sized animals—early mammals, small dinosaurs like Ornithomimids, and even scavenging carcasses left behind by larger predators. His teeth, which would eventually become crushing instruments of bone and flesh, were still developing but already lethal enough to snap through the bones of smaller prey.

His juvenile body, still growing and not yet capable of tackling the Triceratops or Edmontosaurus that roamed the plains, demanded energy from whatever source he could find. Scavenging became an essential skill. Drakon learned to follow the scent of death—whether a kill made by other predators or the result of natural causes. Though not yet the apex predator he was destined to be, young Drakon was a clever opportunist, always on the lookout for a free meal.

On the rare occasion, he ventured into hunting smaller prey, but most of his energy was spent learning the techniques of stealth and strategy. Drakon would crouch low, his brown-green scales blending with the foliage, and wait. Timing was key; a poorly timed attack would leave him without a meal or, worse, vulnerable to larger predators who wouldn't hesitate to see him as prey.

As Drakon grew larger and more confident, he began to set his sights on larger herbivores, stalking the slower, less aware members of herds like Hadrosaurs. Though still far from his full size, his strength was increasing, and his hunting skills were becoming more refined. Drakon was learning to read his environment—the sounds, the smells, the subtle movements that signaled an opportunity.

His future as a fearsome predator was now clear. No longer just a scavenger, Drakon was transforming into a skilled hunter, honing his instincts with every chase and every kill. He was still years away from being the largest and most feared creature in his domain, but the early lessons he learned during his juvenile years would set the stage for his reign at the top of the food chain.

3.2 The Ecosystem and Other Dinosaurs: *Drakon's interaction with other species*

Drakon's world was a rich tapestry of life, teeming with creatures of all shapes and sizes. His juvenile years were spent not just learning how to hunt and survive but also interacting—both directly and indirectly—with the vast array of dinosaurs and other animals that shared his ecosystem.

The forests and plains where Drakon roamed were populated by enormous herbivores, swift raptors, and everything in between. On the open plains, massive herds of Hadrosaurs—duck-billed dinosaurs—grazed peacefully, their long tails swishing back and forth as they tore through the low-lying plants. These herds were a tempting target for any predator, but they moved in large groups for a reason. Together, they were protected from smaller carnivores, and their numbers made it difficult for a young predator like Drakon to isolate a single individual.

Even more imposing were the Triceratops—three-horned giants with large frills shielding their necks. These dinosaurs were formidable not just because of their size, but also their defense mechanisms. They lived in small groups, moving slowly but confidently through the forest and plains. Drakon had learned early on to keep his distance from the adults, whose sharp horns could easily turn a predator into prey. The Triceratops were a constant presence in his world, and while a full-grown Tyrannosaurus would eventually count them as prey, for now, they were an untouchable force in Drakon's environment.

In the skies above, Pteranodons soared gracefully, their leathery wings casting shadows on the ground below. These flying reptiles, while often distant from the terrestrial battles, provided a constant reminder that the airspace, too, was teeming with life. Occasionally, Drakon would spot them diving toward rivers, scooping up fish with their long, toothless beaks, and he knew that the rivers and seas were also filled with creatures vastly different from those on land.

But the real danger in Drakon's world didn't just come from large herbivores. Smaller, faster carnivores like Dromaeosaurs—also known as raptors—were a constant threat. These agile predators moved in packs, their sickle-shaped claws making them formidable foes despite their smaller size. Raptors were known for their intelligence, and while they typically avoided confrontation with a larger predator like Drakon, they were quick to exploit moments of weakness. A juvenile Tyrannosaurus separated from its herd could quickly become prey to a pack of determined raptors.

There were also Alamosaurus, massive sauropods that wandered the plains like living mountains. These long-necked giants were untouchable to all but the fiercest predators, and even then, only the largest adult carnivores dared to attempt a hunt. To young Drakon, they were more of a curiosity than a threat—massive creatures that lumbered slowly through the trees, their huge tails sweeping the ground as they fed on the treetops.

The ecosystem was a delicate balance of predator and prey, a web of life where survival depended on being fast, strong, or smart—or, in the case of the biggest creatures, simply too large to be threatened. As a juvenile Tyrannosaurus, Drakon was still finding his place in this complex world. He was neither the largest nor the fastest predator, but his strength, intelligence, and growing size would eventually place him at the top of this ecosystem.

Drakon also had to navigate interactions with other members of his own species. While adult Tyrannosaurs were known to be solitary hunters, juveniles would sometimes form temporary alliances, working together to scavenge kills or fend off potential threats. These interactions were often tense, as competition for food was fierce, but they provided Drakon with crucial experience in dealing with others of his kind. Eventually, as he grew, he would strike out on his own, but for now, the presence of other juvenile Tyrannosaurs served as both a source of competition and occasional support.

In this vast and vibrant world, Drakon's early experiences shaped him into the predator he would one day become. Every interaction—whether with prey, predator, or his own species—taught him something new about survival in the Late Cretaceous. The world was a dangerous place, filled with creatures each uniquely adapted to their environment. And in this ancient landscape, Drakon was preparing to take his place at the very top.

Quick Facts:

❖ Some dinosaurs, like Velociraptors, may have hunted in packs, while larger predators like T. rex hunted alone.

❖ Dinosaurs like Maiasaura were protective parents, with fossils showing evidence of nesting behavior.

❖ Large herbivores like Hadrosaurs likely traveled in herds, similar to modern-day wildebeest migrations.

Did You Know?

Dinosaurs lived on all continents—even Antarctica!
Fossils of dinosaurs have been found on every continent, including Antarctica, showing that these ancient creatures were truly global.

Chapter 4: The Apex Predator: Drakon in His Prime

4.1 Adult Size and Physical Characteristics

4.2 Hunting Techniques

Chapter 4: The Apex Predator

4.1 Adult Size and Physical Characteristics

By the time Drakon reached adulthood, he had grown into one of the most formidable creatures to ever walk the Earth. His juvenile years, filled with challenges, competition, and learning, had honed his instincts, and now he stood as a true apex predator, ruling over his domain with unmatched strength and dominance.

At full maturity, Drakon was a colossal 12 feet tall at the hips, towering over almost every other creature in his environment. His overall length stretched to an imposing 40 feet, from the tip of his massive head to the end of his thick, muscular tail. This tail, which made up nearly half of his body length, wasn't just for balance—it was a powerful weapon, capable of swinging with enough force to knock smaller creatures off their feet.

Weighing an incredible 9 tons, Drakon's sheer mass alone was enough to intimidate almost any rival or potential prey. His enormous skull, over 5 feet long, was built for power. His jaws, packed with 60 serrated teeth, each up to 8 inches long, were designed to crush bone and tear through the toughest of hides. These teeth weren't just for show—each one was curved backward, making it almost impossible for prey to escape once Drakon had latched on.

Drakon's eyes, set forward in his skull, gave him excellent binocular vision. This allowed him to judge distances with remarkable accuracy, a crucial advantage when stalking prey. His sense of smell was unparalleled—far more acute than that of modern predators like lions or wolves. Drakon could detect the scent of prey from miles away, tracking a wounded dinosaur across vast plains or dense forests.

While his massive head and powerful jaws were his most iconic features, Drakon's body was perfectly designed for his role as a predator. His legs were muscular and sturdy, built to support his enormous weight while allowing him to move with surprising speed when necessary. Though not the fastest dinosaur in his ecosystem, Drakon could reach speeds of 20 miles per hour in short bursts—more than enough to chase down or ambush slower herbivores.

His two-fingered arms, small in proportion to his body, were often seen as a comical feature of the Tyrannosaurus rex. But they weren't useless. These arms, while short, were incredibly strong, capable of holding onto prey or helping him lift himself off the ground if needed. The real power, however, lay in his legs and jaws. Drakon didn't need his arms to dominate his world—his size, strength, and bite were more than enough.

The color of Drakon's skin had also changed over the years. Once a pale brownish-green, perfect for hiding in the dense foliage as a hatchling, his scales had darkened to a deep, earthy green, mottled with streaks of black and brown. This coloration helped him blend into the shadows of the forests or the twilight of the plains, making him a nearly invisible force of nature when stalking his prey.

Drakon was now at the height of his power. He had grown into the largest predator in his ecosystem, unmatched in size, strength, and predatory skill. Other creatures, even the smaller carnivores, would flee at the sight or smell of him, knowing that a confrontation with Drakon often ended in a swift and brutal death.

4.2 Hunting Techniques: *How Drakon hunted and used his physical abilities to survive*

As a fully-grown Tyrannosaurus rex, Drakon was not just powerful—he was also incredibly efficient. Hunting required more than brute strength; it demanded intelligence, strategy, and patience. Drakon had mastered these skills over the years, and his hunting techniques reflected the precision of a top predator.

While many imagine T. rex as a relentless, fast-charging dinosaur that would run down its prey, Drakon preferred to ambush. He was a creature of stealth and strategy. He used his environment to his advantage, blending into the dense forests or crouching low in the tall grasses of the plains. His acute sense of smell and powerful eyesight allowed him to stalk prey from a distance, waiting for the perfect moment to strike.

Drakon primarily targeted large herbivores like Hadrosaurs (duck-billed dinosaurs) and the slow-moving but heavily armored Ankylosaurs. However, his most challenging prey was the Triceratops, the three-horned herbivore that often moved in small herds. Despite their formidable defenses—massive horns, thick frills, and an aggressive attitude—Drakon could still take one down with the right approach.

Patience was key. Drakon would silently observe a herd, looking for signs of weakness—perhaps a juvenile or an injured member lagging behind. He had learned over time that it wasn't always the largest or strongest member of a herd that made for the best meal. Injured or old individuals were easier to bring down, and Drakon's strategy revolved around isolating these weaker members.

Once he had chosen his target, the hunt began. Drakon used his massive tail for balance as he crouched low, waiting for the right moment. When the wind shifted, carrying the scent of the herd toward him, he knew his time had come. With a powerful surge, Drakon would burst from his hiding spot, his legs propelling him forward in long strides. His speed—though limited to short bursts—was enough to close the gap between him and his prey before they even had time to react.

When Drakon reached his target, his jaws would open wide, revealing the fearsome rows of serrated teeth. With a single bite, Drakon could tear through flesh and bone, delivering a fatal wound to his prey. His bite force—one of the strongest ever recorded in the history of life on Earth—was capable of crushing the bones of even the toughest dinosaurs. Unlike other predators, who relied on slashing or tearing, Drakon's bite was designed to break bones, ensuring his prey was too wounded to escape.

Once he had latched onto his prey, there was no escape. His backward-curved teeth made it impossible for any creature to wriggle free. Drakon would use his powerful neck muscles to shake his prey, tearing flesh from bone with each violent movement. Even the mighty Triceratops, with its horns and frill, could not withstand a direct bite from Drakon. One crushing blow to the neck or spine, and the battle was over.

However, Drakon wasn't always about brute force. When necessary, he would scavenge, taking advantage of kills made by other predators or feeding on carcasses left to rot in the hot Cretaceous sun. His keen sense of smell allowed him to detect carrion from miles away, and Drakon was never too proud to feast on a free meal. Scavenging wasn't just a survival tactic—it was part of his strategy for conserving energy. Why risk injury in a fight when a meal could be found with no effort at all?

On rare occasions, Drakon would face competition from other large predators or rival Tyrannosaurs. In these cases, he would use his intimidation tactics—a deep, guttural roar that reverberated across the plains, sending smaller predators running. His size and strength were enough to scare away most rivals, but if a confrontation was unavoidable, Drakon wouldn't hesitate to defend his territory with lethal force.

Drakon's life as an apex predator was one of constant vigilance. Every hunt required skill and strategy, every confrontation tested his dominance. But for now, in his prime, there was no greater predator. Drakon ruled his world with unmatched power, his roars echoing across the Cretaceous landscape like thunder.

The forest trembled beneath his feet. The plains cleared when he appeared. And the once vibrant herds fell silent when they sensed his approach.

This was Drakon's reign—and he ruled it with the fierce might of a true king.

Quick Facts:

- The supercontinent Pangaea began breaking apart during the Mesozoic, creating new ecosystems and isolated dinosaur species.

❖ Volcanic activity both destroyed landscapes and created fertile areas that supported thriving dinosaur populations.

❖ The Mesozoic food chain included plants, herbivorous dinosaurs like Brachiosaurus, and apex predators like T. rex.

Did You Know?
Triceratops had over 800 teeth!
Triceratops didn't just have horns for defense—its powerful jaws were filled with over 800 teeth, perfect for chewing tough plants.

Chapter 5: The Family Life of a Tyrannosaurus

Chapter 5: The Family Life of a Tyrannosaurus

5.1 Social Behavior: *Insights into Drakon's interactions with other Tyrannosaurs*

Drakon, like many apex predators, was often seen as a solitary hunter, the image of a fierce, lone king roaming his domain. But life for a Tyrannosaurus rex wasn't always about isolation. Though his days as a juvenile had been filled with interactions with siblings and other young Tyrannosaurs, as an adult, Drakon still had moments where he encountered others of his kind, and these interactions were as complex and calculated as any hunt.

Despite being primarily solitary, there is evidence that Tyrannosaurs might have had moments of social behavior, especially during the mating season or when disputes over territory arose. Drakon's interactions with other Tyrannosaurs were rare but powerful events, marked by displays of dominance, cooperation, and sometimes conflict.

The world in which Drakon lived was vast, but the territories claimed by each Tyrannosaur were carefully maintained. Boundaries were marked by scent, deep claw marks on trees, and occasionally, by the rotting remains of a kill left as a warning to intruders. When two adult Tyrannosaurs encountered each other, there were no casual greetings. Every meeting was an event, full of tension and the threat of violence. For Drakon, defending his territory meant securing his hunting grounds, protecting his future young, and establishing his dominance in the region.

When he sensed the approach of a rival, Drakon would stand tall, his massive body casting a long shadow across the plains. His roar, deep and thunderous, was his first warning—a declaration of power that echoed across the land. Most of the time, this display of strength was enough. Few creatures in Drakon's world were foolish enough to challenge him, and even other Tyrannosaurs would often retreat rather than face the risk of a battle with an adversary as formidable as Drakon.

But not all encounters ended with a retreat. Occasionally, a younger, less experienced Tyrannosaur would challenge Drakon for dominance. These confrontations were brutal but brief. The two predators would circle each other, roaring and snapping their massive jaws in a display of aggression. Drakon, with his superior size and experience, would often lunge first, using his weight to knock the other Tyrannosaur off balance. With one powerful bite, he could end the challenge and send his opponent fleeing. In rare cases, the fight would escalate into a violent clash, but Drakon's goal was never to kill—simply to maintain his dominance.

But Drakon wasn't always aggressive toward his own kind. During the mating season, his behavior changed, shifting from territorial aggression to a more cooperative and even courting nature. His interaction with females during this time was markedly different, as he focused on securing a mate rather than defending his territory.

5.2 Mating and Reproduction: *Drakon's role in reproduction and guarding the nest*

For Drakon, the instinct to reproduce was as powerful as his instinct to hunt. Every year, during the mating season, the rigid boundaries between Tyrannosaurs softened slightly as the need to pass on their genes took precedence over territorial disputes. Like all animals, reproduction was a crucial part of survival for the species, and Drakon was no exception.

Mating among Tyrannosaurs was a process both dramatic and delicate. Though Drakon was a fearsome predator, his approach to finding a mate required a different set of skills. Tyrannosaur females were just as large and powerful as males, and a clumsy or overly aggressive approach could result in injury or outright rejection.

The courtship process began with vocalizations. Tyrannosaurs, like modern birds (their distant relatives), likely used calls and deep, reverberating roars to signal their presence and readiness to mate. Drakon's roar was not only a declaration of his strength but also an invitation, signaling to any nearby females that he was searching for a mate. The sounds would carry for miles, cutting through the dense forests and open plains.

Once he found a receptive female, Drakon's behavior would shift dramatically. Instead of the aggressive stance he took with rivals, Drakon would engage in ritualistic behaviors designed to demonstrate his strength and fitness without provoking violence. He might circle the female, displaying his massive size and powerful form. The two Tyrannosaurs would engage in what appeared to be a dance—moving around each other, assessing the other's strength and suitability as a mate.

If the female accepted him, mating would occur, and afterward, they would part ways. Unlike some modern animals, Tyrannosaurs did not form long-term pairs. The female would go on to find a safe place to lay her eggs, and Drakon would return to his territory, his role in the reproduction cycle complete—at least for the moment.

However, Drakon's involvement in the reproductive process didn't end with mating. Guarding the nest was a task left primarily to the female, but the male

Tyrannosaurus often stayed close, ensuring that scavengers and other predators didn't threaten the vulnerable eggs.

The nest was an intricate structure, carefully built by the female in a hidden, sheltered area. It consisted of leaves, twigs, and soft earth, formed into a large mound where she would lay her clutch of up to 15 eggs. The eggs themselves were enormous, each about the size of a football, with tough, leathery shells that provided some protection but were still vulnerable to predators.

While the female fiercely guarded the nest, Drakon patrolled the surrounding area, using his size and fearsome reputation to ward off potential threats. His roars were enough to keep most smaller predators at bay, but occasionally, opportunistic scavengers like Dromaeosaurs would try their luck, hoping to steal an egg or two. Drakon's presence during this time was crucial—while the female tended to the eggs, he would chase away any intruders, ensuring that their offspring had a chance to survive.

Once the eggs hatched, Drakon's role became more distant. The hatchlings, while still under their mother's protection, would eventually disperse and begin their own journeys, much like Drakon had in his youth. His duty as a father was more about ensuring the survival of the species as a whole rather than nurturing individual offspring.

In the wild and untamed world of the Cretaceous, life was about survival, and Drakon's legacy was tied to his ability to ensure the continuation of his species. His strength, his territory, and his ability to pass on his genes made him not just a predator, but a crucial part of the ecosystem's balance.

As Drakon roamed his territory, he knew his work was never truly done. Each season brought new challenges—rivals to defeat, prey to hunt, and the cycle of life to continue. And though he was a solitary king in many ways, his impact on the world around him was profound. His offspring, scattered across the plains, carried his genes forward, ensuring that the reign of the Tyrannosaurus rex would continue for generations to come.

In his prime, Drakon had reached the pinnacle of what it meant to be a Tyrannosaurus rex. But the world was changing, and soon, forces beyond his control would shift the balance of life on Earth forever. The age of dinosaurs was nearing its end, and even the mighty Drakon would not be spared.

Quick Facts:

❖ Before paleontology, dinosaur bones were often mistaken for dragons or giants in myth and legend.

❖ Jurassic Park (1993) revolutionized how dinosaurs were depicted in film, using CGI to create lifelike representations.

❖ Dinosaurs are a central part of childhood education and imagination, helping children explore concepts of evolution, extinction, and natural history.

Did You Know?
The ***Stegosaurus*** had a brain the size of a walnut!
Despite its massive body, the Stegosaurus had a very small brain, about the size of a walnut.

Chapter 6: The Fall of the Dinosaurs

6.1 The Extinction Event

6.2 The Struggle for Survival

Chapter 6: The Fall of the Dinosaurs

6.1 The Extinction Event: *The asteroid impact that led to mass extinction*

The world of dinosaurs, which had thrived for over 160 million years, was about to face an unstoppable force—a catastrophe so great that it would wipe out nearly every species on Earth, including the mighty Tyrannosaurus rex. For Drakon, the apex predator of his time, the extinction event would mark the end of his reign and the end of the age of giants.

It began one day, seemingly like any other. Drakon roamed his territory, his powerful legs carrying him across the familiar landscape of dense forests and vast plains. The air was thick with moisture, and the sounds of smaller dinosaurs echoed around him. But unknown to Drakon, a force from beyond Earth's atmosphere was hurtling toward the planet, bringing with it a devastating fate.

An asteroid, estimated to be about 6 miles wide, was on a collision course with Earth. As it sped through space, its impact would forever alter the course of life on this planet. The asteroid struck with unimaginable force in what is now the Yucatán Peninsula in modern-day Mexico, creating the Chicxulub Crater—a scar over 100 miles in diameter. The impact released energy equivalent to billions of nuclear bombs going off simultaneously.

When the asteroid hit, it unleashed shockwaves that rippled through the Earth's crust. Massive tsunamis roared across the oceans, flooding coastlines and sweeping away anything in their path. The impact blasted debris high into the atmosphere, sending molten rock and ash raining down for hundreds of miles, igniting wildfires that spread across the globe. Even far from the point of impact, where Drakon roamed, the consequences were immediate and catastrophic.

Within hours, the skies began to darken. The thick plumes of ash, dust, and gases blocked out the sun, plunging the Earth into a state of near perpetual darkness. This event is often referred to as the "nuclear winter" because the lack of sunlight caused temperatures to drop dramatically. Plants, which relied on sunlight for photosynthesis, began to die off rapidly. Without plants, the herbivores that Drakon relied on for food started to starve.

But it wasn't just the cold and darkness. The atmosphere became toxic, filled with sulfuric acid clouds and noxious gases that turned the rain acidic. Rivers and lakes were polluted, and the very air that Drakon breathed had become tainted. The planet's ecosystem, once vibrant and flourishing, was unraveling at an alarming pace.

For Drakon, this was the beginning of the end. The asteroid impact had set off a chain reaction of destruction that no creature, no matter how powerful, could escape.

6.2 The Struggle for Survival: *Drakon's final days as food sources dwindle*

In the days and weeks that followed the asteroid impact, Drakon's once abundant world became a place of scarcity and desperation. The vibrant ecosystems that had sustained life for millions of years were collapsing. Herds of Hadrosaurs and Triceratops, once so plentiful, had disappeared, either succumbing to the sudden environmental changes or migrating in search of food. Drakon, like all predators, relied on these herbivores for his survival, but now, they were nowhere to be found.

Drakon roamed far beyond his usual hunting grounds, driven by the increasing hunger gnawing at his insides. His once thriving territory was now eerily silent—no rustling of leaves, no distant calls of other dinosaurs, no movement of prey. The thick ash clouds in the sky blotted out the sun, making the days cold and dark. The temperature had dropped, and the lush green forests were slowly withering away, the plants unable to survive without sunlight. Rivers had dried up or become polluted with ash, and even the sounds of insects and birds were becoming faint echoes of the past.

The few animals Drakon encountered were either dying or dead. His acute sense of smell, once his greatest asset in tracking prey, now only led him to the decaying carcasses of dinosaurs that had perished in the days after the impact. Though scavenging had once been a part of his survival strategy, even these sources of food were becoming scarce. The dead had been picked clean by other desperate predators, leaving behind little more than bones.

As the weeks turned into months, the famine deepened. The herbivores, the foundation of the food chain, were gone. Without them, the predators followed. Smaller carnivores, who had once been potential rivals, had either starved or fled in search of food. Even the mighty Triceratops, with their formidable horns and frills, were gone—either killed in the initial impact or unable to survive the environmental collapse.

Drakon, too, was weakening. His massive frame, once so powerful and imposing, now worked against him. His large body required more food and energy than the environment could provide. His once thunderous roar had grown softer, his movements slower and more labored. Every step became a struggle, every breath an effort. The predator who had once ruled the Late Cretaceous with unmatched ferocity was now a shadow of his former self, reduced to scavenging the bones of a world that no longer existed.

The final days of Drakon's life were marked by a deep, instinctual drive to survive, even as the world crumbled around him. His once lush territory, filled with life and sound, was now a barren wasteland. The few remaining trees were blackened by the fires, their branches bare. The ground, once soft and fertile, was hard and cracked beneath his feet. The world he had known had vanished, replaced by a hostile and alien landscape.

In his last days, Drakon was alone. The calls of other Tyrannosaurs, once so familiar, were gone. The vast herds that had roamed the plains had disappeared. The rivers, once teeming with life, were dry. And above it all, the darkened sky, a constant reminder of the catastrophic event that had sealed the fate of his kind.

Drakon's death, like that of so many other dinosaurs, was not dramatic. It came slowly, as his body, weakened by starvation and exhaustion, finally gave out. The predator that had once dominated the Cretaceous, whose roar had echoed across the plains, lay down for the last time. His breath grew shallow, his eyes, once sharp and alert, dimmed. And in the quiet of that final moment, the reign of Drakon, the mighty Tyrannosaurus rex, came to an end.

But while Drakon's individual life had ended, the story of the dinosaurs, including his own, was far from over. Over the millions of years that followed, fossils of Drakon and others like him would be discovered, buried deep beneath the Earth's surface. Paleontologists would study these bones, piecing together the mysteries of a world long gone. Drakon's legacy, along with that of all dinosaurs, would live on—not in flesh, but in the fossilized remains that told the story of the giants who once ruled the Earth.

And so, while the extinction of the dinosaurs marked the end of an era, their memory, their bones, and their story would endure for millions of years, inspiring generations to come.

Did You Know?

Some dinosaurs laid over 20 eggs at a time!

Dinosaurs like ***Maiasaura*** laid clutches of eggs, sometimes as many as 20 to 30 at a time, carefully tending to their nests.

Chapter 7: Legacy and Discovery

7.1 Fossil Discoveries

7.2 The Lasting Impact of Dinosaurs

Chapter 7: Legacy and Discovery

7.1 Fossil Discoveries: How Drakon's species was rediscovered through fossils

For millions of years, Drakon and the other dinosaurs remained buried, their bones entombed beneath layers of rock and earth. The mighty Tyrannosaurus rex had ruled the planet, but like all living creatures, its existence was ultimately finite. The once vibrant, roaring, and hunting Drakon had been reduced to a collection of fossilized bones, silently waiting to be rediscovered by a species that wouldn't walk the Earth until eons later—humans.

The first Tyrannosaurus rex fossils were uncovered in the early 1900s, nearly 66 million years after the last of these great creatures had vanished. The discovery happened in the Hell Creek Formation, an area spanning parts of the western United States, including Montana, Wyoming, and South Dakota. This region, once a lush and fertile environment where dinosaurs roamed, had long since transformed into barren landscapes filled with fossil treasures waiting to be unearthed.

In 1902, a paleontologist named Barnum Brown made a discovery that would change the course of science forever. While excavating in Montana, Brown unearthed the first partial skeleton of a Tyrannosaurus rex. The skeleton wasn't complete, but what Brown found was enough to identify it as a new species, a creature far larger and more fearsome than any previously discovered. The skull alone hinted at a massive predator, with teeth that could crush bone and jaws that could exert thousands of pounds of pressure. Brown's discovery was hailed as a breakthrough, and the world was introduced to the Tyrannosaurus rex—a name that means "Tyrant Lizard King."

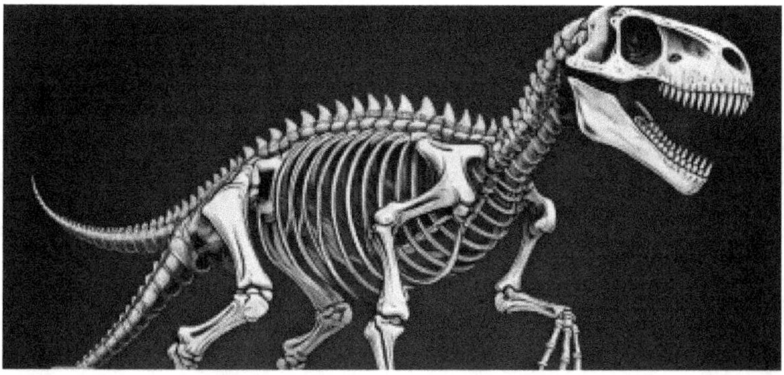

But Brown's discovery was only the beginning. Over the next century, paleontologists would continue to uncover more and more fossilized remains of T. rex, including nearly complete skeletons that revealed new details about the anatomy, behavior, and life of these incredible creatures. Each discovery added another piece to the puzzle of Drakon's world.

Fossils aren't just bones; they are time capsules. Embedded in the rock are not only the remains of the creatures themselves but clues about their environment—the plants they lived among, the creatures they hunted, and the climate in which they thrived. Footprints, teeth marks, and coprolites (fossilized feces) have all helped scientists paint a vivid picture of life during the Late Cretaceous period, when Drakon and his kin roamed the Earth.

In addition to bones, fossilized skin impressions were discovered, revealing what the texture and patterns of a T. rex's skin might have looked like. Though initially thought to be covered in scales, some fossils suggest that juvenile Tyrannosaurs may have had a covering of proto-feathers, a trait shared with their avian descendants, birds. The discovery of these delicate feather imprints revolutionized how we understand the connection between dinosaurs and modern birds.

Drakon, like all dinosaurs, left behind more than just his physical remains. Fossils also provided insight into the structure of his brain, senses, and movement. By scanning T. rex skulls, scientists were able to create 3D models of their brains, revealing that Drakon had exceptionally keen eyesight and an extraordinary sense of smell—confirming his status as one of the most advanced predators of his time.

As paleontologists continue to dig, new discoveries emerge each year. These fossils, preserved in the Earth for millions of years, are still revealing secrets about Drakon's life—how he hunted, how he lived, and ultimately, how his species met its end. The legacy of Drakon, the Tyrannosaurus rex, endures in the bones he left behind, connecting the distant past to the present in ways we are still discovering.

7.2 The Lasting Impact of Dinosaurs: How dinosaurs like Drakon continue to influence modern science and culture

The discovery of Tyrannosaurus rex fossils ignited a fascination with dinosaurs that would captivate the world for generations. From museums to movies, from books to toys, dinosaurs became a symbol of Earth's ancient past, of power, mystery, and extinction. And among them, T. rex stood as the most iconic of all.

One of the greatest impacts dinosaurs like Drakon have had is on the field of paleontology itself. The discovery of dinosaur fossils transformed paleontology into one of the most exciting and respected scientific fields, where researchers continue to unlock the mysteries of Earth's history. The study of dinosaurs has led to profound advancements in evolutionary biology, helping scientists understand how species adapt, survive, and go extinct.

Dinosaurs also helped scientists make the connection between the ancient world and the modern one. Birds, once thought of as entirely separate from dinosaurs, are now recognized as direct descendants of theropods—the group of meat-eating dinosaurs that includes T. rex. This revelation forever changed our understanding of evolution and bridged the gap between two seemingly distant eras of life on Earth.

But beyond science, dinosaurs have had a massive cultural impact. Drakon, the mighty Tyrannosaurus rex, has appeared in literature, movies, and popular culture, symbolizing strength, survival, and the untamed power of nature. From the terrifying portrayal in movies like Jurassic Park to the countless children's books and documentaries, T. rex has become a figure of awe, respect, and wonder.

Museums around the world display the skeletons of T. rex with pride, drawing millions of visitors eager to gaze upon the bones of one of the most fearsome predators to ever live. For many people, seeing a T. rex skeleton is a humbling experience. It's a reminder that the Earth was once home to creatures far more powerful than humans, and that life, in all its forms, is subject to the forces of change, extinction, and evolution.

Dinosaurs, especially T. rex, have also inspired countless generations of children to become scientists, adventurers, and dreamers. The mystery of these ancient creatures sparks curiosity and imagination, pushing young minds to ask questions about the world, about nature, and about the history of life on Earth. The image of a roaring Tyrannosaurus chasing its prey through a primordial forest is one that has been etched into the human psyche, symbolizing both the beauty and brutality of the natural world.

And yet, as much as T. rex represents power and dominance, its extinction serves as a profound reminder of the fragility of life. The asteroid impact that wiped out the dinosaurs shows us that even the mightiest creatures are vulnerable to forces beyond their control. In today's world, this lesson is more important than ever, as humans grapple with the effects of climate change, environmental destruction, and species extinction. The story of the dinosaurs, and Drakon's place in it, is not just a tale of dominance, but also of the delicate balance that all life depends on.

As scientists continue to uncover new fossils and develop new technologies, we are constantly learning more about dinosaurs. The discovery of dinosaur DNA, for instance, has opened up new possibilities for studying these creatures at a molecular level, further deepening our understanding of how they lived, evolved, and died.

In the end, Drakon and his kin live on, not just as fossils in the ground or skeletons in a museum, but as part of the human story. Their legacy is intertwined with our own, reminding us of the ancient past and inspiring us to look forward to the future with curiosity, respect, and wonder.

The roar of Drakon may have been silenced millions of years ago, but its echoes can still be heard today, in every fossil unearthed, in every child's imagination, and in every question scientists ask about the ancient world. His legacy, and that of all dinosaurs, is one that will continue to inspire for generations to come.

Did You Know?

Pterosaurs were not dinosaurs!

Although often confused with dinosaurs, Pterosaurs—flying reptiles—were a separate group of ancient creatures. The largest, Quetzalcoatlus, had a wingspan of nearly 40 feet.

PART-Two

Chapter 8. The Science of Paleontology

8.1 The Process of Fossilization

8.2 Digging for Dinosaurs

8.3 The Evolution of Paleontology

Chapter 8: The Science of Paleontology

Paleontology—the science of uncovering Earth's ancient past—offers us a rare glimpse into a world long gone. The discovery of fossils has transformed our understanding of life on Earth, allowing us to piece together a timeline that stretches back millions of years. Through the study of dinosaurs like **Drakon**, we come to understand not just the creatures themselves but the environments they lived in, how they survived, and, ultimately, how they disappeared. In this chapter, we'll explore the journey of a fossil, from the moment an organism dies to its discovery by modern scientists.

8.1 The Process of Fossilization

Imagine Drakon, the mighty **Tyrannosaurus rex**, meeting his end in the vast plains of the Cretaceous period. Perhaps it was a final confrontation with another T. rex, or maybe the slow decline brought on by the harsh post-asteroid environment. Whatever the cause, his massive body eventually falls to the earth, and with that, the process of **fossilization** begins.

But how exactly does a living creature, full of flesh and blood, turn into stone?

Fossilization is a rare and remarkable event. Most living things, when they die, decompose completely, leaving no trace behind. But under the right conditions, bones, teeth, shells, and even impressions of skin or feathers can be preserved for millions of years. Let's break down the process:

1. **Death and Burial:**

When Drakon dies, his body is exposed to the elements—scavengers, wind, and rain all play their part in breaking it down. However, if Drakon is buried quickly—perhaps by a **mudslide, sandstorm,** or in a **riverbank**—the softer tissues decay more slowly, and his bones are protected from the elements.

1. **Sediment Build-up:**

Over time, **layers of sediment** (mud, sand, ash) accumulate on top of the bones. As the sediment builds up, it compresses and begins to form rock. The bones themselves may also begin to change, as minerals in the surrounding water replace the organic material in the bones.

1. **Permineralization:**

This is the process where the bone is turned to stone. **Minerals**, such as silica or calcite, seep into the tiny spaces within the bones. As these minerals solidify, they slowly replace the original bone material. What's left is a **fossil**, a near-perfect stone replica of Drakon's skeleton.

1. **Erosion and Exposure:**

After millions of years of being buried, the fossilized remains of Drakon are still deep beneath the surface, locked within layers of rock. However, natural forces like **erosion**—caused by wind, rain, or rivers—slowly wear away the surface, eventually revealing the fossil. Sometimes, fossils are buried so deep that they remain hidden for millions of years, until human activity—like digging or construction—accidentally uncovers them.

Not all fossils are bones or teeth. Occasionally, conditions are just right to preserve much more delicate details. **Impressions of feathers, skin texture,** and even **footprints** can survive, giving scientists an incredibly detailed picture of how dinosaurs looked and lived.

8.2 Digging for Dinosaurs

While fossils may remain hidden for millions of years, they are eventually uncovered by the dedicated work of **paleontologists**—scientists who specialize in studying ancient life. But how do paleontologists go about discovering these ancient relics? It's a process that requires patience, precision, and a little bit of luck.

8.2.1 Spotting the Clues Paleontologists don't just stumble upon fossils. They begin their search in areas where rocks from the Mesozoic Era—the age of dinosaurs—are exposed at the surface. This can be in deserts, mountains, or riverbeds, where erosion has worn away the top layers of rock to reveal ancient fossil beds beneath. Sometimes fossils are found by accident during construction projects or even by hikers who notice the tell-tale signs of bone embedded in rock.

When a paleontologist spots what they believe to be a fossil, they approach the find with great care. It might be just a small fragment sticking out of the earth—perhaps a bit of rib, tooth, or vertebra. But this small piece could lead to something much larger.

8.2.2 Excavation: Unearthing the Past Once a potential fossil is discovered, the real work begins: the slow and meticulous process of excavation. Fossils are delicate, and if they are not handled carefully, they can be easily damaged or destroyed. Paleontologists use a variety of tools, from small chisels and brushes to dental picks, to gently remove the surrounding rock (called matrix) without harming the fossil itself.

As they work, paleontologists carefully document the fossil's exact location. Every detail matter because the position of the fossil in the ground, as well as the surrounding sediment, can tell scientists a great deal about the environment in which the dinosaur lived.

Once a fossil is fully excavated, it's often encased in a plaster jacket to protect it during transport. This allows scientists to safely move the fossil to a laboratory, where it can be studied in more detail.

8.2.3 Reconstructing the Ancient World After the fossils are brought to the lab, paleontologists carefully remove the protective plaster and begin cleaning the fossil using fine tools. Once clean, scientists can begin studying the bones—measuring them, scanning them, and even using X-rays or CT scans to look inside.

But excavation is just the beginning. Once the fossil is uncovered, paleontologists face the task of piecing together the skeleton, much like a giant 3D puzzle. The bones of dinosaurs like Drakon are often scattered or incomplete, and paleontologists must use their knowledge of anatomy, combined with comparison to other similar fossils, to reconstruct what the dinosaur looked like.

Sometimes, just a few bones are enough to identify a new species, while other times, nearly complete skeletons are uncovered, offering an almost perfect glimpse into the past.

8.3 The Evolution of Paleontology

The study of fossils may seem like a relatively modern science, but humans have been discovering and marveling at dinosaur bones for centuries—long before the word "dinosaur" even existed.

8.3.1 The Birth of Paleontology In ancient times, people often mistook fossilized dinosaur bones for the remains of giants or dragons. It wasn't until the early 19th century that scientists began to understand that these were the bones of ancient creatures that no longer existed.

The term "dinosaur," meaning "terrible lizard," was coined in 1842 by British paleontologist Sir Richard Owen, marking the formal birth of dinosaur studies. Early paleontologists like Mary Anning, who uncovered many important marine reptile fossils, and Edward Drinker Cope and Othniel Charles Marsh, whose rivalry led to the discovery of hundreds of dinosaur species during the famous "Bone Wars," laid the foundations for modern paleontology.

8.3.2 Modern Advances Today, paleontology has advanced far beyond the early days of bone hunting. Thanks to modern technology, we can now examine fossils in ways that early paleontologists could only dream of.

CT Scanning: By using CT scans, scientists can create detailed, 3D images of the inside of fossils without having to damage them. This allows paleontologists to study the structure of dinosaur bones, including their internal cavities, which can reveal how these creatures moved and breathed.

DNA Research: Although finding dinosaur DNA remains a challenge (DNA degrades over millions of years), advances in genetics are allowing scientists to study ancient species in new ways. By examining the DNA of modern birds (the closest living relatives of dinosaurs), scientists are slowly unravelling the genetic links between dinosaurs and their avian descendants.

Digital Reconstruction: Using digital modelling and 3D printing, scientists can now create incredibly detailed reconstructions of dinosaurs, allowing us to visualize what these creatures might have looked like in life. These models can even simulate muscle movement and behavior, giving us deeper insights into how dinosaurs lived.

From the death of a dinosaur to its rebirth as a fossil, the science of paleontology is a journey of discovery that spans millions of years. Through patience, technology, and a deep curiosity about the past, paleontologists are able to bring creatures like Drakon back to life, piece by piece, fossil by fossil. As we continue to dig into the Earth's past, who knows what new revelations lie buried beneath our feet, waiting to be uncovered?

Did You Know?

The ***T. rex*** could live for up to 30 years.

If a Tyrannosaurus rex survived into adulthood, it could live up to 30 years, ruling its ecosystem for decades.

Chapter 9. Types of Dinosaurs

9.1 Carnivores vs. Herbivores:

9.2 Feathered Dinosaurs:

9.3 Marine and Flying Reptiles:

Chapter 9: Types of Dinosaurs

The age of dinosaurs was a time of astonishing variety, with creatures of every size and shape roaming the Earth. From towering plant-eaters that could stretch their necks into the treetops to fearsome predators with teeth like daggers, each dinosaur was uniquely adapted to survive in the diverse and changing ecosystems of the Mesozoic Era. In this chapter, we'll explore the major categories of dinosaurs—carnivores and herbivores, the growing evidence of feathered dinosaurs, and the equally impressive reptiles that ruled the seas and skies.

9.1 Carnivores vs. Herbivores

The great dinosaur divide was often determined by diet: some dinosaurs were carnivores, preying on other animals, while others were herbivores, feasting on plants. This division shaped every aspect of their lives—from their size and physical features to their behaviors and habitats.

Carnivores: Predators of the Ancient World

Carnivores, like Drakon, the Tyrannosaurus rex, were the apex predators of their time. These meat-eaters had sharp teeth, powerful jaws, and a variety of adaptations that made them expert hunters.

Drakon: Standing at 12 feet tall at the hips and 40 feet long, Drakon was built for power. His massive jaw, filled with 60 serrated teeth, could crush bones and tear through the flesh of his prey. Drakon's size, strength, and speed made him a top predator, capable of taking down large herbivores like Triceratops.

Velociraptors: Unlike the massive T. rex, Velociraptors were smaller, about the size of a turkey, but they were incredibly fast and agile. Their sickle-shaped claws on their feet were perfect for slashing at prey, and they may have hunted in packs, giving them an advantage over larger prey.

While T. rex and Velociraptors are the most famous carnivores, there were many others, each adapted to their specific environment. Some, like Allosaurus, were earlier predators during the Jurassic Period, while others, like Spinosaurus, were uniquely adapted to hunt in both water and on land.

Herbivores: Gentle Giants

At the opposite end of the food chain were the herbivores, the plant-eaters that fed on the rich vegetation of the Mesozoic Era. Some of the largest creatures to ever walk the Earth were herbivores, and they developed their own unique strategies to fend off predators.

Brachiosaurus: Towering over the landscape at nearly 85 feet long and weighing as much as 40 tons, Brachiosaurus was one of the largest land animals to ever exist. Its long neck allowed it to reach the treetops, where it fed on leaves and branches that other herbivores couldn't reach. Despite its enormous size, Brachiosaurus was a peaceful grazer, using its height to access food that was out of reach for most dinosaurs.

Triceratops: In contrast to the long-necked Brachiosaurus, Triceratops was a stocky, four-legged dinosaur with a large frill and three facial horns for protection. At 30 feet long and weighing 6 to 12 tons, Triceratops used its horns to defend itself from predators like Drakon. When threatened, it would charge at its attackers, using its sharp horns to ward off danger.

Herbivores were essential to the ecosystem, converting plant energy into a food source for the carnivores. They evolved a range of defenses to protect themselves, from armored bodies like Ankylosaurus to herding behaviors, which allowed them to band together in numbers for protection.

9.2 Feathered Dinosaurs

While many people imagine dinosaurs as scaly, reptilian creatures, recent discoveries have revealed that some dinosaurs were covered in feathers. This discovery has radically transformed our understanding of these ancient creatures and their relationship to modern birds.

The Feather Revolution

The first hint that dinosaurs might have had feathers came from fossils discovered in China during the late 1990s. These fossils, incredibly well-preserved, showed impressions of feathers surrounding the bones of small, bird-like dinosaurs.

Sinosauropteryx: This small, carnivorous dinosaur, about the size of a cat, was one of the first feathered dinosaurs to be discovered. Its body was covered in simple, hair-like feathers called protofeathers, which were likely used for insulation rather than flight.

Microraptor: Even more remarkable was Microraptor, a small, four-winged dinosaur. This creature had feathered limbs and tail feathers, suggesting that it could glide between trees, much like modern flying squirrels. Microraptor's feathers were iridescent, similar to the feathers of modern crows or magpies, adding a touch of beauty to this ancient predator.

The Bird Connection

Feathered dinosaurs are a crucial link in understanding the evolution of birds. Birds are not just descendants of dinosaurs—they are dinosaurs, part of the same evolutionary family as Tyrannosaurus rex and Velociraptors. The discovery of feathers on non-avian dinosaurs confirms that feathers didn't evolve for flight alone. Instead, they may have originally served purposes like insulation, display, or protection, and only later became adapted for flight in certain species.

Archaeopteryx, often considered the "first bird," lived about 150 million years ago during the Jurassic Period. With a mix of reptilian features (like teeth and a long bony tail) and avian traits (feathers and wings), it provides a perfect snapshot of evolution in action.

The discovery of feathered dinosaurs has given us a more colorful, dynamic view of these ancient creatures. Instead of imagining a world of dull, scaly reptiles, we can now envision a landscape filled with vibrant, feathered creatures, some of which may have sported elaborate plumes for display or camouflage.

9.3 Marine and Flying Reptiles

While dinosaurs ruled the land, other creatures dominated the seas and skies. These weren't technically dinosaurs, but they were reptiles that evolved alongside their land-bound cousins and were just as fearsome.

Marine Reptiles: Masters of the Deep

The oceans of the Mesozoic Era were filled with terrifying marine reptiles that hunted the ancient seas with speed and power.

Mosasaurus: One of the most fearsome predators of the ancient oceans, Mosasaurus was a massive marine reptile, growing up to 50 feet long. With a body built for speed and jaws lined with sharp, conical teeth, Mosasaurus preyed on fish, marine reptiles, and even large sea birds. It was the top predator in the oceans, the aquatic equivalent of T. rex.

Plesiosaurs: Plesiosaurs, with their long necks and broad, paddle-like limbs, were among the most distinctive marine reptiles. They swam gracefully through the water, using their necks to snap up prey like fish or squid. Elasmosaurus, one of the largest plesiosaurs, had a neck that could be over 30 feet long, accounting for nearly half its total body length.

Flying Reptiles: Rulers of the Sky

While the land was dominated by dinosaurs, and the oceans by marine reptiles, the skies belonged to pterosaurs—flying reptiles that evolved to soar above the Mesozoic world.

Pteranodon: One of the most iconic pterosaurs, Pteranodon had a wingspan of up to 33 feet, making it one of the largest flying creatures of all time. Unlike birds, its wings were made of skin and muscle, stretched between its body and a long, slender finger. Pteranodon fed on fish, swooping down to snatch them from the water with its toothless beak.

Quetzalcoatlus: The largest pterosaur ever discovered, Quetzalcoatlus had a wingspan that could reach nearly 40 feet. It likely spent much of its time gliding above the plains, using its height and powerful wings to spot prey or scavenge carcasses. Despite its size, Quetzalcoatlus was lightweight, adapted for long-distance soaring.

These marine and flying reptiles played crucial roles in their ecosystems, ruling their domains as effectively as any land-based dinosaur. Their mastery of the seas and skies completed the triad of dominance during the Mesozoic Era, ensuring that no corner of the Earth was untouched by the reign of reptiles.

The world of dinosaurs was far more diverse and complex than we often imagine. From the giant herbivores like Brachiosaurus to the feathered raptors that glided through forests, and the powerful marine reptiles that patrolled the oceans, the Mesozoic was an era of specialization and evolution. Dinosaurs adapted to every environment on Earth, each species contributing to the incredible mosaic of life during this ancient time. Whether roaming the land, swimming in the seas, or soaring through the skies, these ancient reptiles ruled the planet for over 160 million years, leaving behind a legacy that continues to captivate us today.

Did You Know?

Some dinosaurs swallowed stones to help them digest food.
Herbivorous dinosaurs like ***Brachiosaurus*** would swallow small stones, called gastroliths, to help grind up tough plant material in their stomachs.

Chapter 10: The Evolutionary Link to Modern Animals

10.1 From Dinosaurs to Birds:

10.2 Modern Descendants:

Chapter 10: The Evolutionary Link to Modern Animals

Dinosaurs may have vanished from the Earth 66 million years ago, but their legacy lives on in surprising ways. While the most famous dinosaurs, like Tyrannosaurus rex and Triceratops, are long gone, some of their closest relatives are still with us today, soaring through the skies and lurking in rivers. In this chapter, we will explore how dinosaurs didn't truly die out—they evolved. Their descendants, including modern birds, carry on their legacy in forms we see every day. Additionally, creatures like crocodiles and turtles are living relics from the age of dinosaurs, sharing evolutionary traits with their ancient relatives.

10.1 From Dinosaurs to Birds: Provide a detailed section on how birds evolved from theropod dinosaurs, like Archaeopteryx, and how modern birds carry the legacy of dinosaurs.

Imagine for a moment that the next time you see a sparrow hopping along a branch or an eagle gliding majestically overhead, you're not just watching a bird—you're looking at a living descendant of the dinosaurs. As astonishing as it sounds, birds are direct descendants of theropod dinosaurs, the group that includes famous predators like Tyrannosaurus rex and Velociraptor.

The Archaeopteryx: A Snapshot of Evolution

One of the most significant fossil discoveries in the history of paleontology was that of Archaeopteryx, a small, bird-like dinosaur that lived about 150 million years ago during the Jurassic Period. Archaeopteryx is often described as a "missing link" between dinosaurs and modern birds because it had features of both.

Reptilian Features: Archaeopteryx had sharp teeth, clawed fingers, and a long bony tail, much like its theropod dinosaur ancestors.

Bird-Like Features: But it also had feathers—fully formed feathers on its wings and tail, similar to those of modern birds. These feathers were probably used for gliding or short bursts of flight, making Archaeopteryx one of the earliest examples of flight-capable creatures.

Archaeopteryx and its feathered relatives show us that flight, one of the defining features of birds, didn't appear overnight. It evolved gradually from non-flying theropod dinosaurs, who first used feathers for insulation and display, and only later adapted them for gliding and flying.

Feathers: More Than Just for Flight

Feathers, once thought to be unique to birds, were actually present on many dinosaurs. Fossil evidence shows that even Velociraptor, a swift predator, had feathers. This discovery has reshaped our understanding of how feathers evolved. Initially, feathers likely served purposes other than flight, such as:

Insulation: Keeping smaller dinosaurs warm in cold environments.

Display: Just like modern peacocks, some dinosaurs may have used colorful feathers to attract mates or intimidate rivals.

Camouflage: Feathers may have helped smaller dinosaurs blend into their surroundings to avoid predators.

As time passed, these feathered theropods evolved the musculature and bone structure needed for flight, giving rise to the first true birds. These small, feathered creatures gradually became better adapted to life in the trees and skies, evolving into the diverse range of birds we see today.

Modern Birds: The Living Dinosaurs

Today, the world is home to over 10,000 species of birds, and every one of them is a direct descendant of theropod dinosaurs. From the smallest hummingbird to the mighty albatross, birds retain many traits of their ancient relatives.

Hollow Bones: One of the most important adaptations for flight is the presence of hollow bones—a feature that reduces weight while maintaining strength. Theropod dinosaurs also had hollow bones, a trait passed down to modern birds.

Three-Toed Feet: Like their dinosaur ancestors, most birds have three forward-pointing toes on each foot, a characteristic that evolved from their theropod predecessors.

Wishbone (Furcula): The wishbone or furcula, which strengthens the chest for flight muscles, is a feature that first appeared in theropods and has been carried forward into birds.

Even behaviors like nest building and caring for young, once thought to be unique to birds, are now believed to have existed in some dinosaurs. Fossils have revealed dinosaur nests and even evidence of dinosaurs sitting on their eggs, much like birds do today.

So, the next time you see a bird perched on a branch, know that you're witnessing the result of millions of years of evolution—a lineage that stretches back to the mighty dinosaurs of the Mesozoic Era.

10.2 Modern Descendants: Show how creatures today—like crocodiles, turtles, and birds—share evolutionary traits with the dinosaurs of the Mesozoic Era.

Birds aren't the only creatures alive today that can trace their ancestry back to the age of dinosaurs. Other modern reptiles, like crocodiles, alligators, and turtles, are part of a group that survived the mass extinction at the end of the Cretaceous period. These animals share several evolutionary traits with their ancient ancestors, making them some of the oldest living creatures on Earth.

Crocodiles: Living Relics from the Mesozoic

Crocodiles are often called "living fossils" because they've remained relatively unchanged for millions of years. In fact, their ancestors lived alongside dinosaurs, sharing the same environments and ecosystems.

Crocodyliforms: During the time of the dinosaurs, there existed a diverse group of reptiles known as crocodyliforms. Some of these creatures were as long as 40 feet and lived in both freshwater and marine environments. They were formidable predators, capable of hunting large prey, including dinosaurs.

Survival through Adaptation: One reason crocodile survived the mass extinction event that wiped out the dinosaurs is their ability to thrive in aquatic environments. When the asteroid hit, causing widespread fires and climate changes, many terrestrial animals died out. Crocodiles, however, could retreat to rivers and lakes, where they could find food and shelter from the worst of the environmental changes.

Today's crocodiles retain many of the traits that helped their ancestors survive, such as their powerful jaws, strong swimming abilities, and tough, armored skin. While they are no longer the massive creatures of the Mesozoic, they remain one of the most successful predators in the animal kingdom.

Turtles: Survivors from an Ancient Lineage

Like crocodiles, turtles have been around since the time of the dinosaurs, with the first turtles appearing more than 200 million years ago.

Ancient Origins: The earliest turtles didn't have the hard, protective shells we associate with them today. Over time, their shells evolved to provide protection from predators, making them one of the most successful groups of reptiles in Earth's history.

Evolutionary Success: Turtles are incredibly adaptable creatures, having survived through numerous environmental changes, including the asteroid impact that wiped out the dinosaurs. Their ability to live in a variety of habitats—from freshwater rivers and lakes to the open ocean—helped them endure through the mass extinction events.

Today, turtles can be found in nearly every part of the world, from desert tortoises to sea turtles that traverse vast distances across oceans. They serve as a living reminder of Earth's ancient past, carrying with them traits that have remained relatively unchanged for hundreds of millions of years.

Other Modern Survivors

In addition to birds, crocodiles, and turtles, there are several other species that carry the evolutionary legacy of the Mesozoic:

Lizards and Snakes: These reptiles share common ancestors with dinosaurs, and while they evolved separately, they still exhibit traits that connect them to the ancient world.

Amphibians: Creatures like frogs and salamanders have ancestors that lived during the time of the dinosaurs. Their ability to adapt to both water and land environments helped them survive the mass extinction.

Though the age of dinosaurs came to a dramatic end 66 million years ago, their story didn't end with the asteroid impact. It continues to this day through the birds that soar above us and the reptiles that swim in our rivers and oceans. These modern creatures carry the genetic and evolutionary legacy of the Mesozoic Era, reminding us that even after the greatest extinction events, life finds a way to survive and evolve.

From the mighty Tyrannosaurus rex to the sparrows chirping in our backyards, the line of life stretches across eons, linking the ancient past to the present day. Dinosaurs may have vanished, but their descendants are still all around us, living reminders of Earth's distant and dynamic history.

Did You Know?

Ankylosaurus had built-in armor.

The *Ankylosaurus* was like a walking tank, covered in bony armor plates and armed with a massive clubbed tail to defend itself against predators.

Chapter 11: Dinosaur Behavior and Social Structures

11.1 Pack Hunters and Solitary Predators:

11.2 Dinosaur Parenting:

11.3 Migration and Herding:

Chapter 11: Dinosaur Behavior and Social Structures

Dinosaurs were not just massive, mysterious creatures that roamed the Earth—they were also social beings, with complex behaviors and social structures. Though much of what we know about dinosaur behavior comes from fossil evidence, modern technology and discoveries have given us remarkable insights into how these ancient creatures lived, hunted, raised their young, and even travelled in herds. In this chapter, we'll dive into the pack hunting of carnivores, the parenting instincts of some species, and the herding and migration patterns of the massive plant-eaters that once walked the Earth.

11.1 Pack Hunters and Solitary Predators

The Mesozoic landscape was teeming with predators, each equipped with unique tools to survive. But how these predators hunted—whether alone or in groups—varied greatly, depending on their size, intelligence, and prey. Fossil evidence offers clues about which dinosaurs may have hunted in packs, like modern wolves, and which ones were solitary, relying on brute force to secure a meal.

Pack Hunting: Velociraptors and Small Carnivores

When people think of Velociraptors, they often envision the swift, intelligent hunters from Jurassic Park—a vision not far from the truth, though the real Velociraptor was smaller, about the size of a turkey. Fossil evidence, including tracks and bones found together, suggests that some smaller carnivores, like Velociraptors and Deinonychus, may have hunted in groups, or at least engaged in some form of cooperative behavior.

Velociraptor fossils have been discovered alongside the remains of larger prey, like Protoceratops, indicating that these swift predators may have worked together to bring down animals larger than themselves. Their long, sickle-shaped claws were perfect for slashing at prey, and hunting in packs would have given them a strategic advantage.

Deinonychus, a slightly larger cousin of Velociraptor, is often considered the best candidate for pack hunting based on fossil evidence. Several fossils have been found near the skeletons of large herbivores, suggesting these carnivores may have worked in coordinated groups, attacking from multiple sides.

Pack hunting isn't a common strategy in the animal kingdom, but when it is employed, it's highly effective. Working together, pack hunters could take down much larger prey than they could on their own, ensuring that everyone in the group gets a share of the kill. This type of hunting requires a degree of intelligence and social coordination, traits that some theropods may have possessed.

Solitary Predators: Tyrannosaurus Rex

On the other hand, Drakon, the Tyrannosaurus rex, was likely a solitary predator. At a towering 40 feet long and weighing 9 tons, T. rex didn't need a pack to hunt. His sheer size, power, and ability to ambush prey made him a force to be reckoned with.

T. rex's behavior is a subject of debate among paleontologists, but most agree that this massive carnivore didn't rely on pack strategies. Fossils suggest that T. rex likely hunted large herbivores like Triceratops or Hadrosaurs on its own, using its bone-crushing bite to disable prey quickly. T. rex's hunting method may have involved ambush tactics, using its acute sense of smell to track prey from miles away before attacking with overwhelming force.

There is also some evidence, based on fossilized bite marks, that T. rex may have been a scavenger at times, using its strength to drive other predators away from kills. Solitary predators often have to balance between hunting and scavenging, depending on what opportunities arise in their environment.

In the world of dinosaurs, hunting strategies varied widely, but whether hunting in packs or on their own, carnivores were some of the most efficient predators to ever walk the Earth.

11.2 Dinosaur Parenting

Dinosaurs weren't just savage predators and prey—they were also parents who cared for their young. Fossil evidence has revealed much about the parenting behavior of certain species, showing that some dinosaurs, like modern birds, built nests, protected their eggs, and even cared for their hatchlings.

Maiasaura: The Good Mother Lizard

One of the most famous examples of dinosaur parenting comes from Maiasaura, whose name literally means "Good Mother Lizard." Fossils of this medium-sized herbivore, which lived about 75 million years ago, were discovered in Montana in the late 1970s. Paleontologists unearthed an extraordinary find: a nesting site containing eggs, hatchlings, juveniles, and adults, all in the same area. This evidence showed that Maiasaura lived in herds and that the adults stayed close to their young after they hatched.

Nests: Maiasaura nests were large, shallow depressions dug into the ground and lined with vegetation to protect the eggs. These nests were spaced closely together, much like a colony of modern seabirds.

Caring for Hatchlings: Fossils of baby Maiasaura show that the hatchlings weren't capable of walking right away. This suggests that the adults brought food to the nest, feeding their young until they were strong enough to walk and fend for themselves. The close proximity of nests suggests group protection of the colony, with adults likely warding off predators.

Oviraptors: Not Just Egg Thieves

For many years, Oviraptors were wrongly believed to be egg thieves, due to the discovery of an Oviraptor fossil near a clutch of eggs. However, further studies revealed that the eggs didn't belong to another dinosaur—they were Oviraptor eggs, meaning the adult was likely protecting its nest when it died, not stealing eggs. This discovery changed the perception of Oviraptors, showing that they were likely attentive parents.

Oviraptor fossils, including those of adults found sitting on nests, have given us an incredible look at dinosaur nesting behavior. Like birds today, Oviraptors may have kept their eggs warm by brooding, using their bodies to incubate the eggs until they hatched.

11.3 Migration and Herding

The Mesozoic Era wasn't just home to solitary predators and small family groups. It was also an era of massive herds of herbivores, some of which likely engaged in seasonal migration, traveling long distances to find food and water. These herds included some of the largest animals ever to walk the Earth.

Hadrosaurs: The Duck-Billed Dinosaurs

Hadrosaurs, often called duck-billed dinosaurs due to the shape of their snouts, were some of the most successful herbivores of the Late Cretaceous. Fossil evidence, including trackways showing hundreds of footprints moving in the same direction, suggests that Hadrosaurs lived in herds and may have migrated seasonally in search of better grazing lands, much like modern-day bison or wildebeest.

Fossilized Trackways: These long trails of footprints give us a snapshot of life during the Cretaceous. In some areas, scientists have found tracks that show entire herds of Hadrosaurs moving together, likely in search of new feeding grounds as the seasons changed. These herds may have included thousands of individuals, moving across vast distances in search of food.

Protection in Numbers: Living in herds gave Hadrosaurs a significant advantage over predators. While a single Hadrosaur might have been vulnerable to attack from a pack of raptors or a solitary T. rex, a large herd could ward off predators through sheer numbers. The safety of the herd would have been particularly important for juveniles, who would have stayed near the center of the group for protection.

Sauropods: Giants on the Move

Even larger than Hadrosaurs were the Sauropods—massive, long-necked dinosaurs like Apatosaurus and Brachiosaurus. These dinosaurs likely travelled in herds, too, moving slowly across the landscape as they stripped entire forests of vegetation. Fossils of sauropod nesting grounds suggest that they returned to the same areas year after year to lay their eggs, much like modern sea turtles.

Nesting Behavior: Like other herd animals, Sauropods likely gathered at specific nesting sites to lay their eggs. Fossilized nests have been found in clusters, suggesting that these giant dinosaurs may have returned to the same locations annually to reproduce, laying eggs and then moving on.

Migration for Food: Given the enormous appetites of Sauropods, it's likely that they had to move constantly to find enough food to sustain their massive bodies. A single Brachiosaurus could eat hundreds of pounds of vegetation a day, and once a forest was stripped bare, the herd would have had no choice but to move on to greener pastures.

Dinosaur behavior and social structures were as varied and complex as the creatures themselves. From pack hunters like Velociraptors to solitary predators like Drakon, dinosaurs developed unique strategies to survive in a world filled with both danger and opportunity. The parenting instincts of species like Maiasaura reveal that dinosaurs weren't just cold-blooded reptiles—they were capable of nurturing and protecting their young, just like modern birds and mammals. And the herding behavior of giant herbivores shows that these creatures weren't just wandering aimlessly—they moved with purpose, migrating in search of food and protection.

Through the fossil record, we're able to catch glimpses of how these incredible animals lived, hunted, raised their young, and travelled across the Earth. Though the dinosaurs are long gone, their behaviors still echo in the lives of animals today, from birds brooding over their nests to herds of elephants migrating across the African savannah. In many ways, the world of dinosaurs is still with us.

Did You Know?

Some dinosaurs may have been warm-blooded.

Recent studies suggest that some dinosaurs, especially smaller, feathered species, might have been warm-blooded, like modern birds.

Chapter 12: Climate and Ecosystem of the Mesozoic Era

Chapter 5: Climate and Ecosystem of the Mesozoic Era

The Mesozoic Era—known as the "Age of Reptiles"—was a time of tremendous change on Earth. During this period, dinosaurs rose to dominate the planet, adapting to shifting continents, volcanic landscapes, and a complex food web that supported an incredibly diverse range of life. In this chapter, we will explore how the continents broke apart, how volcanic activity shaped the world, and how dinosaurs fit into the intricate food chain that supported life on Earth for millions of years.

12.1 Changing Continents

The Mesozoic Era, which lasted from approximately 252 to 66 million years ago, saw the breakup of the supercontinent Pangaea, one of the most significant geological events in Earth's history. This massive landmass began to break apart during the Jurassic Period, and by the Cretaceous Period, the continents were drifting into the positions that would eventually form the Earth's modern layout. The shifting of continents had profound effects on the climate, ecosystems, and dinosaur habitats.

The Supercontinent Pangaea

At the beginning of the Triassic Period, all of Earth's landmasses were joined together in one vast supercontinent called Pangaea. This immense landmass stretched from pole to pole and was surrounded by the vast ocean, known as Panthalassa. During this time, much of the Earth was arid, with large desert-like regions. Dinosaurs, having recently evolved, began to spread across this single landmass.

The central regions of Pangaea were hot and dry, while the coasts enjoyed more temperate climates due to their proximity to the ocean. This uniformity of land and climate allowed dinosaurs to spread and diversify rapidly.

Breaking Apart: The Birth of Oceans and New Habitats

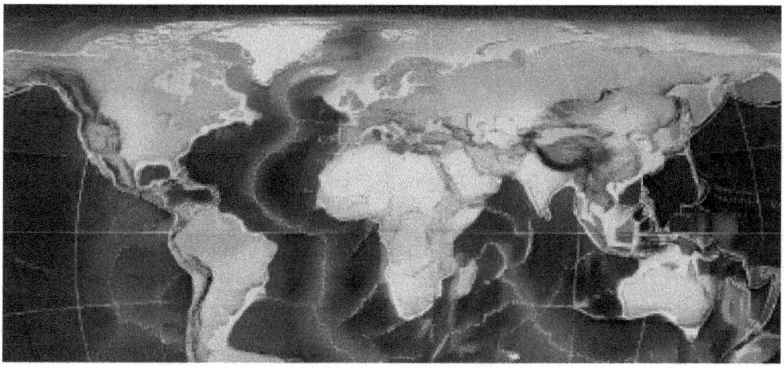

Around 200 million years ago, during the Jurassic Period, Pangaea began to break apart, driven by the slow but powerful movements of tectonic plates. The giant landmass split into two large supercontinents: Laurasia in the north and Gondwana in the south.

Laurasia: Comprised of what would become North America, Europe, and Asia.

Gondwana: Included the landmasses that would become South America, Africa, Antarctica, Australia, and India.

As these landmasses continued to drift, new oceans and seas began to form between them. This breakup created isolated ecosystems, forcing dinosaurs to adapt to different climates and environments. Dinosaurs living in what would become North America evolved separately from those in South America or Africa, leading to an explosion of new species.

The separation of landmasses also affected climates. Coastal regions experienced wetter, more humid conditions, while inland areas remained dry. These new climates created rainforests, swamps, and plains—all ideal environments for different species of dinosaurs. For instance, large sauropods thrived in lush forested areas, while the mighty Tyrannosaurus rex roamed the open plains of North America during the Late Cretaceous.

The breakup of Pangaea also led to the development of mountain ranges, river systems, and coastal environments, all of which helped shape the biodiversity of the Mesozoic.

12.2 The Role of Volcanoes

Volcanic activity played a significant role in shaping the Mesozoic landscape. Volcanoes, fueled by the movement of tectonic plates, were incredibly active during this era, influencing the climate, the geography of the land, and the survival of the creatures living on Earth.

Volcanoes and the Mesozoic Environment

As the continents drifted apart, the movement of tectonic plates created immense volcanic ranges. Volcanic eruptions released massive amounts of lava, ash, and gases into the atmosphere, which could have both positive and negative effects on life during the Mesozoic.

Ash Clouds: Volcanic eruptions would sometimes release large clouds of ash into the atmosphere, temporarily cooling the planet by blocking out sunlight. This cooling effect could lead to the extinction of plant species in affected areas, which in turn impacted the herbivores that relied on these plants for food.

Lava Flows: Lava flows from volcanic eruptions could reshape entire landscapes, destroying forests and rivers but also creating new geological features, such as mountains and fertile plains.

The Fertile Ash

While volcanic eruptions could be destructive, they also created fertile soils rich in minerals, which promoted plant growth. After the initial devastation, these fertile areas allowed plants to repopulate quickly, creating lush forests and abundant food sources for herbivorous dinosaurs. In fact, many of the most verdant habitats where dinosaurs thrived were the result of volcanic soils.

This balance between destruction and renewal created dynamic ecosystems where dinosaurs could thrive. While large eruptions may have occasionally disrupted ecosystems, they also helped renew the land, providing fertile ground for the plants that sustained vast populations of herbivores, which in turn supported predators like Drakon, the Tyrannosaurus rex.

12.3 Dinosaur Food Chain

The Mesozoic Era was a time of incredible diversity, with ecosystems filled with plants, herbivores, carnivores, and scavengers. Dinosaurs occupied nearly every part of the food chain, from small insect-eating creatures to massive predators like T. rex. Understanding this food web is essential to understanding how life flourished during the Mesozoic.

Producers: Plants

At the base of the food chain were the plants, which used sunlight to grow and provided the primary food source for herbivores. During the Mesozoic, gymnosperms (non-flowering plants like ferns, cycads, and conifers) dominated the landscape, providing food for many species of herbivorous dinosaurs. By the Cretaceous period, flowering plants (angiosperms) began to emerge, changing the types of vegetation available.

Ferns and Conifers: These plants were the primary food sources for large herbivores, such as Brachiosaurus and Stegosaurus.

Angiosperms: The rise of flowering plants during the Cretaceous provided new food sources for smaller herbivores, contributing to a greater diversity of plant-eating dinosaurs.

Primary Consumers: Herbivores

The herbivores of the Mesozoic era were some of the largest creatures to ever walk the Earth, consuming vast amounts of plant matter to fuel their enormous bodies.

Sauropods: Massive, long-necked dinosaurs like Apatosaurus and Diplodocus grazed on the tops of trees, using their long necks to reach the highest branches.

Hadrosaurs: Also known as duck-billed dinosaurs, these herbivores roamed the forests and plains in large herds, grazing on low-lying plants and leaves.

Ceratopsians: Dinosaurs like Triceratops, with their impressive horns and frills, used their powerful beaks to shear through tough vegetation, grazing on ferns and cycads.

These herbivores, in turn, supported a wide variety of carnivores, creating a balanced and interconnected food web.

Secondary Consumers: Carnivores

At the top of the food chain were the carnivores, which fed on the herbivores and smaller animals. These predators varied in size and hunting methods, from small pack hunters like Velociraptors to solitary giants like Tyrannosaurus rex.

Small Carnivores: Dinosaurs like Compsognathus and Troodon hunted insects, small mammals, and smaller dinosaurs, filling a crucial niche in the food web.

Medium-Sized Predators: Dromaeosaurs (like Velociraptor) hunted in packs, using speed, intelligence, and cooperation to bring down prey larger than themselves.

Apex Predators: Dinosaurs like T. rex were the ultimate hunters, using their size, strength, and ferocious bite to dominate the ecosystems they lived in.

Scavengers and Decomposers

Not every dinosaur was a predator or prey—some species filled the essential role of scavengers, feeding on the remains left behind by other animals. Large carnivores like T. rex may have scavenged when the opportunity arose, feeding on carcasses they encountered. Small carnivores and early mammals also scavenged, playing a vital role in recycling nutrients back into the ecosystem.

Additionally, decomposers like bacteria, fungi, and insects helped break down dead plant and animal matter, enriching the soil and allowing the cycle of life to continue.

The Mesozoic Era was a time of immense transformation. As Pangaea broke apart and the landmasses drifted into the shapes we recognize today, dinosaurs adapted to the changing climates and environments. Volcanic activity

reshaped the world, creating both challenges and opportunities for life to flourish. And within these dynamic landscapes, dinosaurs filled every niche in the food chain, from the smallest insect-eating species to the largest herbivores and apex predators.

The ecosystems of the Mesozoic were rich and varied, supporting an incredible diversity of life. From the towering sauropods grazing on treetops to the ferocious Tyrannosaurus rex stalking the plains, each dinosaur played a crucial role in the food web that sustained life on Earth for millions of years.

Did You Know?

Some dinosaurs may have been warm-blooded.
Recent studies suggest that some dinosaurs, especially smaller, feathered species, might have been warm-blooded, like modern birds.

Chapter 13: Dinosaurs and Human Culture

13.1 Dinosaurs in Myth and Legend

13.2 Dinosaurs in Pop Culture

13.3 The Dinosaur Craze

Chapter 13: Dinosaurs and Human Culture

For millions of years, dinosaurs lay buried beneath the Earth, their bones untouched and their existence unknown. But once these ancient creatures were unearthed, they captured the human imagination in ways few scientific discoveries ever have. From the myths and legends of ancient peoples to their rise as stars in popular culture, dinosaurs have left an indelible mark on how we view the past—and ourselves. In this chapter, we'll explore the powerful influence of dinosaurs on human culture, from the legends they inspired to their starring roles in books, films, and childhood fantasies.

13.1 Dinosaurs in Myth and Legend

Long before the first fossils were studied by scientists, ancient peoples often encountered strange and massive bones embedded in rock or scattered across landscapes. With no concept of extinct creatures, these early societies created myths and legends to explain the origins of these bones. It's no surprise that tales of dragons, giants, and other mythical creatures have been linked to the discovery of dinosaur remains.

Dragons: A Legacy of Dinosaur Bones

In many cultures, dragons are depicted as large, fearsome reptiles with massive bodies, long tails, and sharp teeth—strikingly similar to what we now know about some dinosaurs. This connection between dragons and dinosaurs is likely no coincidence. In China, for example, fossilized dinosaur bones were sometimes thought to be the remains of dragons—creatures believed to hold both great power and wisdom.

Ancient China: In Chinese tradition, dragons were revered as symbols of power, strength, and good fortune. It's possible that early Chinese farmers and miners uncovered fossilized remains of large dinosaurs, which fueled their stories of mythical beasts. Even today, some rural communities in China refer to dinosaur bones as "dragon bones."

Medieval Europe: In Europe, similar legends emerged. Fossilized bones of large creatures, including Megaloceros (giant elk) and even dinosaurs, were sometimes mistaken for the remains of giants or dragons. These discoveries may have contributed to the tales of fire-breathing dragons that terrorized villages and were ultimately slain by knights.

Giants in Mythology

In other parts of the world, the discovery of large bones led to stories of giants—humanoid creatures that once walked the Earth. In Greek mythology, the Cyclopes and Titans were thought to be massive, ancient beings. The discovery of large, fossilized bones likely reinforced these beliefs. For instance, fossils of woolly mammoths or dinosaurs might have been interpreted as the remains of these giants.

Native American Legends: Some Native American tribes also created legends around the discovery of large bones. For example, the Sioux believed that the Thunderbirds, large bird-like creatures with immense power, were responsible for shaping the land. These myths might have been inspired by the discovery of Pterosaur fossils.

Long before paleontology emerged as a scientific discipline, dinosaur bones were woven into the fabric of human storytelling. These bones became symbols of mystery and wonder, inspiring some of the greatest legends and myths of human history.

13.2 Dinosaurs in Pop Culture

While ancient peoples interpreted dinosaur bones through the lens of mythology, modern society has embraced dinosaurs in an entirely different way. Over the past two centuries, dinosaurs have gone from scientific curiosities to global cultural icons, appearing in everything from books and movies to toys and video games. Dinosaurs hold a unique place in pop culture, often symbolizing the balance between science and imagination.

Early Representations in Art and Literature

Dinosaurs first made their mark on the popular imagination in the mid-19th century, soon after the term "dinosaur" was coined in 1842 by British paleontologist Sir Richard Owen. Early depictions of dinosaurs in art, such as the life-sized models created for the Crystal Palace exhibition in London in the 1850s, were speculative and often inaccurate, but they captivated audiences nonetheless.

As scientific discoveries of dinosaurs expanded, artists and writers began imagining these creatures in various ways. Early 20th-century adventure novels, such as Arthur Conan Doyle's The Lost World (1912), depicted dinosaurs as fearsome beasts that survived in remote corners of the Earth, waiting to be discovered by daring explorers.

The Rise of Dinosaurs in Film: From Stop-Motion to CGI

The true explosion of dinosaurs into pop culture came with the advent of film. Early monster movies, like The Lost World (1925), used stop-motion animation to bring dinosaurs to life on the big screen. In King Kong (1933), dinosaurs appeared alongside the giant ape, battling it out in dramatic sequences that thrilled audiences.

But it was Steven Spielberg's Jurassic Park (1993) that redefined how the world saw dinosaurs. Using ground-breaking computer-generated imagery (CGI) and animatronics, the film portrayed dinosaurs in a way that had never been done before. The T. rex, Velociraptors, and other dinosaurs in Jurassic Park were terrifyingly lifelike, their movements fluid and realistic. Audiences were stunned by the spectacle of these ancient creatures walking, running, and hunting in front of their eyes.

Jurassic Park not only revolutionized film technology, but it also reignited a worldwide obsession with dinosaurs. The movie series continues to this day, with sequels like Jurassic World (2015) captivating new generations of fans. The franchise has helped shape the public's perception of dinosaurs, with the Velociraptor and Tyrannosaurus rex becoming household names.

Dinosaurs in Art and Media

Beyond films, dinosaurs have appeared in countless forms of media, from animated TV shows like The Land Before Time (1988) to children's books, toys, and even theme parks. Dinosaurs are used in advertisements, museum exhibitions, and educational programs, making them one of the most recognizable symbols in the world.

Dinosaurs have become so ingrained in pop culture that they often serve as metaphors for power, extinction, or danger, whether in cartoons, comic books, or even political cartoons. Their imagery is synonymous with both awe and fear, reminding us of Earth's distant past and the forces that shaped its history.

13.3 The Dinosaur Craze

Today, dinosaurs are as much a part of childhood as they are a part of science. From an early age, children are fascinated by these giant creatures, collecting dinosaur toys, watching dinosaur cartoons, and learning the names of species that roamed the Earth millions of years ago. This dinosaur craze shows no sign of slowing down, and it plays an important role in both education and imagination.

Dinosaurs in Childhood Education

Dinosaurs are often one of the first topics children encounter when learning about science and natural history. Their incredible size, strange shapes, and mysterious extinction spark a natural curiosity in young minds. Dinosaurs provide a gateway to learning about:

Geology: Through dinosaurs, children are introduced to concepts like fossils, rock layers, and the history of Earth.

Evolution: Dinosaurs help explain the process of evolution, showing how life on Earth changes over millions of years.

Extinction: The story of the dinosaurs' extinction helps children understand the delicate balance of ecosystems and the potential threats to modern species.

Dinosaurs are featured in countless children's books, ranging from the purely educational to the wildly imaginative. Books like *Dinosaur Train*, *How Do Dinosaurs Say Goodnight?*, and National Geographic Kids: Dinosaurs offer kids a fun yet informative way to engage with these ancient creatures.

The Power of Imagination

Dinosaurs also play a huge role in children's creative play. With their imposing size, varied appearances, and fantastic stories, dinosaurs encourage kids to engage in role-playing, imaginative storytelling, and exploration. From digging up plastic bones in sandboxes to constructing dinosaur-themed LEGO sets, kids can act out the adventures of their favorite species.

The fascination with dinosaurs also extends into video games, where players can control, fight, or even raise dinosaurs in various fantasy and simulation worlds. Games like Ark: Survival Evolved and Jurassic World Evolution allow players to interact with dinosaurs in a fully immersive environment.

For children, dinosaurs offer a perfect blend of education and wonder, helping to develop scientific thinking while fueling the imagination. This early fascination often sparks a lifelong interest in paleontology, natural history, and the ancient world.

Dinosaurs have captured the hearts and minds of humans for centuries, evolving from mysterious creatures of myth to beloved stars of modern pop culture. Whether they are inspiring tales of dragons and giants, thrilling moviegoers in blockbuster films, or becoming a fixture of childhood education, dinosaurs have a unique ability to bridge the gap between science and storytelling. They are timeless symbols of life's power and fragility, reminding us of the grandeur of Earth's history and our ongoing fascination with the unknown.

From the ancient myths of dragons to the CGI dinosaurs that dominate movie screens, these incredible creatures remain as relevant and captivating today as they were when they first walked the Earth millions of years ago. Their story isn't just one of extinction—it's one of eternal curiosity, imagination, and discovery.

Did You Know?

Dinosaurs and humans never coexisted.
Although they dominate pop culture, dinosaurs went extinct about 66 million years ago, long before humans appeared.

Chapter 14: Deeper Scientific Insights for Older Audiences

(The Debate on Dinosaur Extinction)

14.1 Beyond the Asteroid

14.2 Could Dinosaurs Come Back?

Chapter 14: Deeper Scientific Insights for Older Audiences

As readers grow older and delve deeper into the scientific aspects of dinosaurs, more complex topics—like the debate on dinosaur extinction and the possibility of resurrecting ancient species—become fascinating subjects of exploration. In this section, we will present the scientific arguments and theories that continue to intrigue paleontologists and geneticists today.

14.1 Beyond the Asteroid: The Debate on Dinosaur Extinction

For many years, the leading theory behind the extinction of the dinosaurs has been the asteroid impact, which struck the Earth about 66 million years ago. However, while the asteroid theory is widely accepted, it's now believed that other contributing factors, such as volcanic activity and climate change, played critical roles in the mass extinction event.

The Asteroid Impact

The asteroid, believed to have been around 6 miles wide, slammed into what is now the Yucatán Peninsula in Mexico, forming the Chicxulub Crater. The impact released an unimaginable amount of energy—equal to billions of atomic bombs—and triggered catastrophic events worldwide.

Global Fires and Tsunamis: The immediate effects of the asteroid included massive wildfires, tsunamis, and a shockwave that reverberated across the planet.

Dust and Ash Cloud: Perhaps more devastating was the cloud of dust, soot, and aerosols that rose into the atmosphere, blocking out sunlight for months or even years. This event, often called a "nuclear winter," led to a dramatic drop in global temperatures and disrupted photosynthesis, causing a collapse of food chains.

The sudden lack of sunlight likely caused many plant species to die off, leading to the collapse of ecosystems that relied on those plants. Herbivorous dinosaurs would have starved, followed by the predators that hunted them.

Volcanic Activity: The Deccan Traps

While the asteroid caused immediate devastation, scientists have increasingly turned their attention to volcanic activity in a region known as the Deccan Traps (in present-day India) as another major contributor to the extinction.

Massive Eruptions: The Deccan Traps were experiencing one of the largest volcanic events in Earth's history during the same period, with massive eruptions that lasted for thousands of years. These eruptions spewed lava, ash, and sulfur dioxide into the atmosphere, which could have had long-term effects on the climate.

Cooling and Acid Rain: The release of sulfuric gases would have led to acid rain and global cooling, compounding the effects of the asteroid impact. These eruptions might have weakened ecosystems even before the asteroid struck, contributing to the long-term decline of many species.

Climate Change: A Gradual Shift

In addition to the asteroid and volcanic activity, climate change was already underway during the Cretaceous Period. For millions of years, the climate had been slowly cooling, and sea levels had been fluctuating, which may have altered ecosystems and habitats.

Changing Ecosystems: As sea levels fell and rose, coastal habitats were transformed, and inland regions experienced drier conditions. This gradual shift in the climate may have put additional stress on dinosaur populations, particularly those living in areas that became less hospitable over time.

The combination of these events—climate shifts, volcanic eruptions, and the asteroid impact—created a perfect storm that led to the mass extinction known as the Cretaceous-Paleogene (K-Pg) extinction event. While the asteroid is often seen as the final blow, it's likely that these other factors played significant roles in the decline and eventual extinction of the dinosaurs.

14.2 Could Dinosaurs Come Back?

For decades, the idea of resurrecting dinosaurs has fascinated scientists and the public alike, largely thanks to films like Jurassic Park. While the science fiction concept of cloning dinosaurs from preserved DNA seems exciting, the reality of bringing dinosaurs back is far more complex. That said, genetic engineering is making strides, leading some scientists to ask: Could dinosaurs really come back?

The Problem of Dinosaur DNA

The key challenge in resurrecting dinosaurs is DNA degradation. DNA is a fragile molecule that breaks down over time, especially after millions of years. No dinosaur DNA has been found in fossils, and it's unlikely that DNA could survive for the 66 million years since dinosaurs went extinct.

Fossilization: During the fossilization process, soft tissues like DNA degrade and are replaced by minerals, making it extremely unlikely that any recoverable genetic material remains in dinosaur fossils.

Ancient DNA Limits: So far, the oldest DNA recovered from a fossil belongs to a 1.2-million-year-old mammoth found in the Arctic. This is far younger than the dinosaurs, showing the extreme difficulty of finding usable DNA in older specimens.

The "De-Extinction" Approach: Reverse Engineering

While cloning a dinosaur from fossilized DNA seems unlikely, scientists are exploring the possibility of using genetic engineering to recreate certain dinosaur traits in modern animals—specifically, birds.

Birds as Dinosaur Descendants: Since modern birds are direct descendants of theropod dinosaurs, some scientists believe they could use CRISPR and other gene-editing technologies to reverse-engineer certain dinosaur characteristics.

This process would involve modifying bird genomes to "switch on" dormant genetic traits from their dinosaur ancestors.

The Chickenosaurus Project: One famous example is the Chickenosaurus Project, led by paleontologist Jack Horner. This experiment aims to reverse-engineer a chicken to express dinosaur-like traits, such as a long tail, clawed hands, and teeth. By tweaking certain genes, researchers hope to create a creature that resembles a small dinosaur.

Ethical and Ecological Considerations

Even if it became scientifically possible to recreate dinosaurs, there are significant ethical and ecological concerns to consider.

Ecosystem Impact: Modern ecosystems are not designed to accommodate dinosaurs. Introducing dinosaurs back into the wild could have unpredictable effects on the environment and existing species. How would dinosaurs interact with today's predators or prey, and what role would they play in a modern ecosystem?

Animal Welfare: There are also concerns about the well-being of any resurrected dinosaurs. Would these creatures suffer in environments that are drastically different from their native Cretaceous world? Without the right habitats, they could struggle to survive, raising questions about whether it's ethical to bring them back at all.

Fiction Meets Reality: What's Next?

While the Jurassic Park scenario remains highly unlikely, the possibility of recreating dinosaur traits through genetic engineering offers a tantalizing glimpse into what the future might hold. For now, "de-extinction" efforts are more focused on species that went extinct more recently, like the woolly mammoth. Researchers are actively working on cloning these creatures using preserved DNA from frozen specimens.

As genetic engineering and cloning technologies advance, it's possible that we may one day see animals with dinosaur-like traits roaming the Earth once again.

However, bringing back actual dinosaurs remains a scientific challenge that will likely remain in the realm of science fiction for the foreseeable future.

Summary

The extinction of the dinosaurs is one of the most fascinating and debated topics in science, with new discoveries and theories continuing to emerge. While the asteroid impact is the most famous explanation, factors like volcanic eruptions and climate change also played vital roles in their downfall. At the same time, the idea of resurrecting dinosaurs through genetic engineering captures our imaginations, even though the science remains in its infancy.

As we continue to explore these deeper scientific questions, it's clear that dinosaurs will forever remain a source of wonder, mystery, and curiosity.

Chapter 15: Advances in Dinosaur Research

Chapter 15: Advances in Dinosaur Research

As technology advances, so does our ability to unlock the secrets of the ancient world, including dinosaurs. Paleontology has evolved from simply digging up bones to using state-of-the-art technologies that allow scientists to reconstruct dinosaur behavior, understand their biology, and even explore the tantalizing possibility of recovering dinosaur DNA. In this chapter, we will explore some of the most exciting modern techniques that are revolutionizing the study of dinosaurs.

15.1 Technology in Paleontology

In the past, paleontologists relied on traditional methods of discovery—excavation, cleaning, and studying fossils. While these methods remain critical, modern technology has provided new tools that allow scientists to dive deeper into the past, offering unprecedented insights into how dinosaurs lived, moved, and even thought.

CT Scanning: Looking Inside Fossils

CT scanning, or computed tomography, is one of the most revolutionary technologies used in paleontology today. CT scanners allow paleontologists to look inside fossils without damaging them, revealing detailed structures that were once hidden.

Bone Structure: By using CT scans, paleontologists can examine the internal structure of dinosaur bones, including growth patterns, blood vessel channels, and hollow spaces. For example, scanning the skull of a Tyrannosaurus rex can show the arrangement of its braincase and how its senses—like smell or sight—might have worked.

Brain Reconstruction: With CT scans, scientists can create 3D models of dinosaur brains, allowing them to understand more about how these creatures processed information. For example, scans of Velociraptor skulls have revealed enlarged regions in the brain that control balance and coordination, suggesting these predators were highly agile.

Muscle Attachment: Scans of fossilized bones can also reveal muscle attachment sites, helping scientists estimate the size and power of the muscles that dinosaurs used to move. This is essential for reconstructing how dinosaurs walked, ran, or used their limbs for hunting.

3D Printing: Bringing Dinosaurs to Life

3D printing has become another game-changer in paleontology. Once fossils are scanned, paleontologists can create highly accurate, to-scale 3D models of dinosaur bones. This allows for detailed study and even the reconstruction of entire skeletons.

Replica Fossils: With 3D printing, scientists can create replicas of delicate fossils that are too fragile to handle directly. These replicas allow researchers to experiment with different hypotheses about movement, behavior, or structure without risking the original fossils.

Rebuilding Skeletons: By printing individual bones, scientists can piece together full skeletons of species that are incomplete in the fossil record. These models help scientists and museum visitors alike understand the full size and posture of dinosaurs.

Simulating Movement: 3D-printed models can be used to test how dinosaurs might have moved. For example, by examining the joint articulation in 3D-printed Velociraptor limbs, paleontologists can simulate how these predators leapt or slashed with their claws.

Virtual Reality and Digital Modeling

Virtual reality (VR) and digital modeling have introduced new ways to study dinosaurs. By combining CT scans and 3D models, paleontologists can build interactive digital simulations of dinosaurs and their environments.

Movement Simulations: Digital models of dinosaur skeletons are used to simulate their movements. By analyzing the structure of limbs, muscles, and joints, researchers can predict how dinosaurs walked, ran, or even hunted. For instance, T. rex simulations show that it was likely a fast, but not overly agile, predator due to its massive size.

Behavioral Insights: VR allows paleontologists to place their digital models in reconstructed prehistoric environments. This can help scientists understand how dinosaurs interacted with their surroundings, what their hunting behaviors may have been like, or how they might have cared for their young.

These technological advances not only allow scientists to better understand how dinosaurs lived but also make the field of paleontology more accessible to the public through digital reconstructions and museum exhibits that bring dinosaurs to life in ways never before possible.

15.2 The Hunt for Dinosaur DNA

Ever since the idea was popularized by films like Jurassic Park, scientists and the public alike have wondered whether it's possible to recover dinosaur DNA and, perhaps one day, clone or genetically modify living creatures to resemble them. While this remains a long shot, there have been some intriguing discoveries and advances that fuel ongoing scientific debate.

Challenges in Recovering Dinosaur DNA

One of the major challenges in recovering dinosaur DNA is that DNA degrades over time. Even in ideal conditions, DNA can only survive for a few hundred thousand years, far less than the 66 million years since dinosaurs went extinct.

No Intact DNA Found: Despite years of searching, no intact dinosaur DNA has been discovered. DNA is a fragile molecule, and over millions of years, it breaks down due to environmental factors like heat, moisture, and radiation.

Soft Tissue Discoveries: However, in recent years, paleontologists have made exciting discoveries, such as soft tissues in well-preserved fossils. In some cases, traces of proteins like collagen have been found, giving hope that researchers might one day recover fragments of DNA. For example, scientists studying a fossilized T. rex leg bone found remnants of soft tissue, an extraordinary find that could lead to further breakthroughs.

The Limits of De-Extinction Science

While cloning dinosaurs from ancient DNA is unlikely, scientists have begun to experiment with the idea of de-extinction—the process of using genetic engineering to bring back extinct species or create hybrids that resemble extinct creatures. However, there are significant challenges to this idea.

Birds as Dinosaurs' Living Relatives: Since birds are the direct descendants of theropod dinosaurs, some scientists have explored the possibility of using reverse engineering to recreate certain dinosaur traits in birds. By altering the genetic code of modern birds, scientists might be able to "turn on" dormant dinosaur-like traits, such as teeth, long tails, or clawed hands.

The Chickenosaurus Project: This experiment, led by paleontologist Jack Horner, seeks to modify the genome of chickens to express ancestral traits found in dinosaurs. While this wouldn't technically bring back dinosaurs, it could create a bird with more dinosaur-like characteristics. However, this project is still in its early stages, and many ethical and technical questions remain.

The Future of Ancient DNA Research

Despite the limitations, the search for ancient DNA is far from over. New technologies, such as gene editing (using tools like CRISPR), are opening doors to what might be possible in the future.

Mammoth Resurrection: While no one has found dinosaur DNA, scientists have been more successful with recently extinct species, such as the woolly mammoth. Using DNA recovered from frozen mammoth remains, researchers are now working to clone or genetically modify modern elephants to express mammoth-like traits, such as fur and adaptations for cold climates.

Environmental DNA (eDNA): Researchers are also experimenting with eDNA, which is genetic material found in the environment (like soil or water) rather than in fossils. This could offer new avenues for discovering ancient genetic traces, though it's unlikely to lead to full dinosaur genomes.

The future of paleogenetics may be focused on recent extinctions, but as technology improves, scientists could learn more about ancient species like dinosaurs by studying the genes of their closest living relatives.

Summary

Advances in technology have transformed the field of paleontology, allowing scientists to study dinosaurs in ways that were previously unimaginable. From CT scans and 3D printing to virtual reality reconstructions, these tools are unlocking new insights into how dinosaurs lived, moved, and interacted with their environments. Meanwhile, the search for dinosaur DNA, though still elusive, continues to drive the dreams of scientists and enthusiasts alike.

As our understanding of genetics and ancient biology improves, we may come closer to recreating parts of the dinosaur world—even if it's through modified birds or other creative applications of genetic engineering. Though bringing dinosaurs back remains a far-off goal, the technological strides being made today are helping us better understand these ancient creatures than ever before.

Did You Know?

The asteroid that killed the dinosaurs was six miles wide.

The asteroid that caused the extinction of the dinosaurs 66 million years ago was about six miles wide and hit Earth with the force of billions of atomic bombs.

Chapter 16: The Future of Dinosaur Research

Chapter 16: The Future of Dinosaur Research

Dinosaurs may have roamed the Earth millions of years ago, but our fascination with them is stronger than ever. With every new discovery, scientists are uncovering more about how these ancient creatures lived, evolved, and disappeared. The study of dinosaurs has expanded far beyond mere curiosity—it's become a vital tool for understanding the history of life on Earth and the challenges we face in the modern world. In this conclusion, we will explore the future of dinosaur research and its broader implications for science and society.

16.1 Ongoing Discoveries

Paleontology is not a static field. Each year, new fossil discoveries are made, leading to significant breakthroughs in our understanding of the past. New species of dinosaurs are being unearthed, providing fresh insights into the incredible diversity of life during the Mesozoic Era.

New Fossils and Technologies

Thanks to modern excavation techniques, many new fossils are discovered in areas previously thought to be unremarkable. Some of the most groundbreaking finds have occurred in places like China's Liaoning Province, where fossils of feathered dinosaurs have completely transformed our understanding of the connection between dinosaurs and birds.

Advances in technology, such as CT scans, 3D modeling, and geochemical analysis, are allowing paleontologists to study fossils in ways that were unimaginable just a few decades ago. These tools enable scientists to:

1. Reconstruct muscle and tissue structures to better understand dinosaur movement.
2. Analyze growth rings in bones to learn about dinosaur lifespans and growth rates.
3. Examine isotopes in fossils to deduce dinosaurs' diets and habitats.
4. As fossil-hunting techniques improve and new sites are explored, it's likely that the coming decades will bring even more astonishing finds, further illuminating the mysterious and ancient world of dinosaurs.

16.2 Connecting Dinosaurs to Modern Conservation

The extinction of the dinosaurs, caused by a massive asteroid impact and a series of volcanic eruptions, wiped out nearly 75% of all life on Earth. By studying this mass extinction event, scientists gain critical insights into the fragile nature of ecosystems and the potential threats that face species today.

Learning from Extinction

Dinosaurs' extinction teaches us about the interconnectedness of life on Earth. When a catastrophic event, such as the asteroid impact, dramatically altered the climate and food chains, it set off a domino effect that led to the collapse of ecosystems. This has strong parallels to the challenges facing modern species, including:

Climate change: Just as drastic environmental shifts affected the dinosaurs, today's climate change threatens ecosystems worldwide, with rising temperatures, shifting weather patterns, and habitat loss endangering many species.

Habitat destruction: Dinosaurs thrived in a wide range of environments, but as their habitats changed dramatically, many species were unable to adapt. Today, the destruction of habitats through deforestation and urbanization is leading to the extinction of numerous species.

Biodiversity loss: The extinction of dinosaurs led to the loss of countless species and altered the course of evolution. Similarly, modern species face extinction at an alarming rate due to human activity, causing irreversible changes to ecosystems.

By studying the end of the dinosaurs, scientists hope to better understand how species and ecosystems respond to massive environmental changes. This knowledge can be used to inform modern conservation efforts and help protect endangered species from the threats of climate change and habitat destruction.

The Sixth Extinction?

Some scientists argue that we are currently in the midst of a sixth mass extinction, driven primarily by human activity. While the dinosaurs' extinction was caused by natural disasters, today's mass extinction is being fueled by deforestation, pollution, climate change, and overexploitation of natural resources. Understanding the past can help us address these challenges and develop strategies to preserve the diversity of life on Earth.

16.3 Final Reflection

Dinosaurs offer us a unique glimpse into Earth's deep history, a time when creatures that seem almost mythical walked the planet. They remind us that life is ever-changing and that the world we live in is shaped by forces beyond our control, from shifting continents to asteroid impacts. Yet, they also inspire us to continue exploring, discovering, and learning.

A Reminder of Earth's History

Dinosaurs are a reminder of Earth's long and complex history—a history filled with dramatic changes in climate, geography, and life itself. The rise and fall of these ancient reptiles show us that nothing on Earth is permanent. Species evolve, adapt, and eventually, many disappear. While dinosaurs are long gone, they continue to teach us about our own place in the world and the importance of understanding the natural cycles of life and extinction.

Our Connection to the Natural World

In studying dinosaurs, we also strengthen our connection to the natural world. Their fossils offer tangible evidence of life that existed long before humans, and their extinction offers important lessons about environmental stewardship. As we face challenges like climate change and biodiversity loss, dinosaurs remind us of the importance of preserving the delicate balance of life on Earth.

A Journey of Discovery

For readers, the study of dinosaurs is not just about the past—it's an invitation to embark on a journey of discovery. Whether you're fascinated by the T. rex's powerful bite, the mystery of the feathered Velociraptor, or the incredible diversity of herbivores like Triceratops, dinosaurs spark a sense of wonder that transcends time. They inspire curiosity about the world we live in and encourage us to keep asking questions, digging deeper, and seeking answers.

As you finish this exploration into the world of dinosaurs, remember that the story of life on Earth is far from over. Paleontology is an ever-evolving science, and new discoveries will continue to enrich our understanding of the ancient past. The lessons we learn from dinosaurs can guide us in shaping a better future for the species that share our planet today. The legacy of dinosaurs lives on—not just in their fossils, but in our continued quest for knowledge and understanding.

The End... Or Just the Beginning?

The journey of discovering dinosaurs is never truly over. With each fossil unearthed, we uncover more pieces of the puzzle that is Earth's ancient history. So, whether you're a budding paleontologist or simply a dinosaur enthusiast, remember—there's always more to learn, more to explore, and more mysteries waiting to be uncovered. Dinosaurs may be extinct, but their legacy is very much alive.

Did You Know?

Dinosaurs weren't the only creatures of the Mesozoic Era.
While dinosaurs ruled the land, Pterosaurs ruled the skies, and marine reptiles like Mosasaurus dominated the oceans.

Epilogue: Roar of the Ages

Millions of years have passed since Drakon, the mighty Tyrannosaurus rex, took his last steps across the Cretaceous plains. His reign as the most fearsome predator of his time has long since ended, but the story of dinosaurs—of creatures like Drakon—lives on. These ancient titans, whose roars once echoed through dense forests and across vast, open landscapes, left a legacy that continues to shape our understanding of life on Earth.

For centuries, the bones of these great creatures lay buried, locked away beneath layers of earth and rock, waiting for their story to be told. And when humans finally uncovered these fossils, we began piecing together a narrative far older and grander than we could have ever imagined. Through the study of dinosaurs like Drakon, we were given a window into a world long gone—a world of giants, predators, and ecosystems so complex that they rival anything we see today.

But the legacy of dinosaurs is not just about the past; it's also about the future. The discovery of fossils changed how we view our planet's history and sparked a curiosity that would lead to groundbreaking scientific advancements. Dinosaurs have become more than just relics of an ancient world—they are symbols of life's resilience, evolution, and the delicate balance that sustains all living things.

In museums around the world, the skeletons of Tyrannosaurus rex stand proudly, towering over visitors and capturing the imaginations of people from all walks of life. Children gaze in awe, their minds filled with images of Drakon stalking the plains, his powerful tail swaying, his sharp eyes locked on his next meal. Adults reflect on the grandeur of Earth's natural history, marveling at how such creatures once roamed the same planet we now inhabit.

Dinosaurs, especially the mighty T. rex, remind us of the vastness of time. They walked the Earth for millions of years, far longer than humans have existed, yet they were not immune to the forces of nature. Their extinction, sudden and absolute, serves as a humbling reminder that no species, no matter how powerful, is invincible. The asteroid impact that ended their reign reminds us of the fragility of life, of how quickly and unpredictably the balance of nature can shift.

But despite their extinction, dinosaurs never truly left us. Their bones have become timeless messengers, offering clues about Earth's ancient climate, the evolution of species, and the intricate web of life that once thrived. As we continue to study them, we learn not only about the past but also about our present and future—about how species adapt, how ecosystems evolve, and how life can endure through the most extreme challenges.

Dinosaurs like Drakon also serve as powerful symbols in our culture. They embody both the awe-inspiring power of nature and its ultimate vulnerability. Whether in films, books, or art, dinosaurs have become iconic figures, representing not just a world that once was, but the mysteries that still remain. Their presence in our imagination is a testament to humanity's innate curiosity and desire to understand the world around us.

The story of Drakon, and all dinosaurs, is one of survival, dominance, and extinction, but it is also one of discovery, learning, and wonder. From the moment their bones were first unearthed, dinosaurs have captured the human spirit, pushing us to explore, to ask questions, and to seek knowledge about the distant past and the unknown future.

As we continue to dig deeper into the Earth, both literally and metaphorically, we uncover more of their story. New fossils, new technologies, and new theories allow us to refine our understanding of these magnificent creatures. In a way, dinosaurs are still very much alive—not in flesh, but in the profound impact they have on science, education, and culture.

Drakon, like his fellow dinosaurs, will forever remain a part of Earth's story. His roar, though silenced by the forces of nature millions of years ago, still reverberates through time, echoing in every fossil discovery, in every classroom where children learn about the ancient world, and in every scientific breakthrough that brings us closer to understanding life's great mysteries.

As we reflect on the Roar of the Ages, we remember that the story of life on Earth is one of constant change—of rise and fall, of creation and extinction. Dinosaurs may no longer roam the Earth, but their legacy continues to shape our understanding of the natural world. And in that sense, their roar will never truly fade.

Did You Know?

Dinosaurs likely had good eyesight and sense of smell.

Predators like T. rex had excellent eyesight and a sharp sense of smell, helping them track prey over long distances.

Glossary

Carnivore: An animal that eats meat.

Cretaceous Period: The final period of the Mesozoic Era, lasting from about 145 to 66 million years ago, when many famous dinosaurs, including Tyrannosaurus rex, roamed the Earth.

CT Scan: A medical imaging technique that allows scientists to see inside fossils without damaging them.

Dinosaur: A group of reptiles that lived during the Mesozoic Era, characterized by their upright stance.

Extinction: The permanent disappearance of a species.

Fossil: The remains or impression of a prehistoric organism preserved in rock.

Fossilization: The process by which a plant or animal becomes a fossil, usually through mineral replacement over millions of years.

Herbivore: An animal that eats plants.

Jurassic Period: The middle period of the Mesozoic Era, lasting from about 201 to 145 million years ago, when dinosaurs like Brachiosaurus and Stegosaurus lived.

Mesozoic Era: The age of the dinosaurs, lasting from about 252 to 66 million years ago, and divided into the Triassic, Jurassic, and Cretaceous periods.

Omnivore: An animal that eats both plants and animals.

Paleontologist: A scientist who studies fossils and ancient life.

Paleontology: The scientific study of life in the geological past, primarily through the analysis of fossils.

Theropod: A group of bipedal, mostly carnivorous dinosaurs, including Tyrannosaurus rex and Velociraptor, that are closely related to birds.

Triceratops: A large, herbivorous dinosaur from the Cretaceous period known for its three horns and large bony frill.

Did You Know?

Some dinosaurs had incredible horns and frills.
The ***Styracosaurus*** had a frill lined with six long horns, making it one of the most fearsome-looking herbivores of its time.

Further Reading and Resources

For readers who want to dive deeper into the fascinating world of dinosaurs and paleontology, this section will provide a curated list of books, documentaries, and websites. Here are some recommendations:

Books:

The Rise and Fall of the Dinosaurs by Steve Brusatte

A comprehensive and engaging overview of dinosaur evolution and their eventual extinction, written by a leading paleontologist.

Dinosaurs: A Very Short Introduction by David Norman

An accessible introduction to dinosaur science, covering key concepts and recent discoveries.

Dinosaur Encyclopedia by DK Publishing

A beautifully illustrated, kid-friendly resource with in-depth facts about a wide range of dinosaur species.

Documentaries:

Walking with Dinosaurs (BBC)

A groundbreaking series that brings dinosaurs to life with cutting-edge CGI and detailed reconstructions of their world.

Dinosaur 13 (2014)

A documentary that tells the fascinating and controversial story of the discovery of the largest T. rex fossil, known as "Sue."

Prehistoric Planet (Apple TV+)

A recent nature documentary series narrated by David Attenborough that showcases dinosaurs in incredible, lifelike detail using the latest in CGI technology.

Websites:

National Geographic: Dinosaurs

[www.nationalgeographic.com/
dinosaurs](https://www.nationalgeographic.com/animals/prehistoric/
dinosaurs)

Offers articles, photos, and videos exploring the latest discoveries and facts about dinosaurs.

The Paleontology Portal

www.paleoportal.org

A great resource for students and teachers, covering a wide range of topics on fossils, paleontology careers, and recent news.

Fossilworks

www.fossilworks.org

A database that provides access to fossil data from around the world, perfect for the scientifically curious.

Did You Know?
The tallest dinosaurs were taller than a five-story building.
Some sauropods, like Brachiosaurus, could reach heights of over 40 feet, making them taller than most buildings!

Bibliography

Books:

Brusatte, Steve. *The Rise and Fall of the Dinosaurs*: A New History of a Lost World. William Morrow, 2018.

- Norman, David. *Dinosaurs: A Very Short Introduction*. Oxford University Press, 2005.

Academic Papers:

- Horner, Jack. "*Evidence of Dinosaur Egg Brooding Behavior in the Cretaceous of Montana*." Science, vol. 259, no. 5096, 1993, pp. 790-793.

- Norell, Mark A., et al. "*A Theropod Dinosaur Embryo and the Affinities of the Flaming Cliffs Dinosaur Eggs*." Science, vol. 266, no. 5186, 1994, pp. 779-782.

Websites:

- National Geographic, "*Why Did Dinosaurs Go Extinct?*"

www.nationalgeographic.com/science/article/why-did-dinosaurs-go-extinct

Smithsonian National Museum of Natural History, "*Dinosaurs: Fossils and Paleobiology*"

naturalhistory.si.edu/research/paleobiology/dinosaurs

Other Books By the Author

- Raja Porus: An Unsung Hero of Punjab
- Behind the Crown: The Hidden Life of Alexander the Great
- Paws, Wings and Hearts: Understanding and Bonding with Your Pets
- Echoes of the Asian Elephant: A Journey from Majesty to Survival
- Behind Tomorrow: The World in 2125
- The Psychology of Sensual Love: Understanding Obsession and its Consequences
- Introduction to Digital Communication Skills
- The Reality of Ghairat (Honor)
- Mastering Prompt Engineering: Unlocking AI's Full Potential
- The Art of Detecting Lies: Techniques for Home, School, and Workplace

Publisher: Muhammad Yoosuf Bhutta at ***Gulbang Publishers***

Printer: Muhammad Yoosuf Bhutta at ***Gulbang Printers***

Place of publishing: H. No. B-II/320, Al-karam Street, Jalalpur Road,

Gujrat-50700. Pakistan

Place of printing: Chenab Plaza, Court Road, Gujrat-50700. Pakistan

Edition: First

Price: US $ 16.99

Email: yoosufbhutta@yahoo.com

Did You Know?

The longest dinosaur name is "Micro-pachy-cephalo-saurus." This tiny dinosaur had a name that's almost longer than its body! Micro-pachy-cephalo-saurus means "tiny thick-headed lizard."

Also by Yoosuf Bhutta

Love Beyond Borders: True Stories of International Romance
Love Beyond Borders: True Stories of International Romance

Standalone
Beyond Tomorrow: The World in 2125
Roar of the Ages: The Life Story of Drakon, the Mighty Tyrannosaurus
Introduction to Digital Communication Skills

About the Author

About The Author

Muhammad Yoosuf Bhutta is a distinguished journalist, writer, and educator with a career spanning over four decades in the fields of journalism and communication. With an M.Sc. in Mass Communication (graduating with distinction), a PGD in Television Production, and an impressive portfolio of hundreds of published articles, features, and translations in both Urdu and English, he has established himself as a versatile and insightful writer.

Throughout his career, Bhutta has taken on diverse roles, including correspondent, sub-editor, feature writer, and magazine editor, showcasing his ability to navigate complex topics with analytical depth and clarity. His passion for knowledge extends beyond the newsroom into academia, where he has spent over eight years as a visiting lecturer in communication and business studies at university-affiliated campuses. Here, he has inspired students to think critically and engage with the evolving dynamics of the media landscape.

With additional academic accomplishments such as an MBA in Marketing, a BS in Applied Psychology, and 17 professional certifications across multiple disciplines, Bhutta brings a wealth of knowledge and a unique interdisciplinary perspective to his writing.

In his latest eBook, **Introduction to Digital Communication Skills**, Bhutta leverages his extensive expertise to demystify the complexities of the digital communication era. Drawing on decades of experience in traditional and digital media, he guides readers through the transformative technologies and strategies that define modern communication. His work not only equips readers with practical knowledge but also inspires them to appreciate the profound ways digital communication is reshaping our world. Bhutta's ability to weave technical insights with real-world examples ensures this eBook is both accessible and impactful for students, professionals, and anyone curious about navigating the digital age.

www.ingramcontent.com/pod-product-compliance
Lightning Source LLC
LaVergne TN
LVHW022110200926
848481LV00003B/1139
* 9 7 9 8 2 3 0 3 8 6 5 6 8 *